THE APOLLO 11 MOON LANDING

40th Anniversary Photographic Retrospective

Compiled By

DENNIS R. JENKINS and **JORGE R. FRANK**

specialtypress
PUBLISHERS AND WHOLESALERS

SPecialty press
PUBLISHERS AND WHOLESALERS

Specialty Press
39966 Grand Avenue
North Branch, MN 55056 USA
Phone: (651) 277-1400 or (800) 895-4585
Fax: 651-277-1203
www.specialtypress.com

ISBN 978-1-58007-148-2
Item Number SP148

Library of Congress Cataloging-in-Publication Data

Jenkins, Dennis R.
  Apollo 11 / by Dennis R. Jenkins & Jorge R. Frank.
     p. cm.
  ISBN 978-1-58007-148-2
  1. Project Apollo (U.S.)--History. 2. Space flight to the moon--History. I. Frank, Jorge R. II. Title.
  TL789.8.U6A51175 2008
  629.45'4--dc22
                                    2008034470

Printed in China
10 9 8 7 6 5 4 3 2 1

Special Thanks to ...

Neil Armstrong and Mike Collins, each of whom graciously offered comments to a draft copy of this publication; Ron Stephano, who stitched together many of the panoramic photos shown here; Barbara Green and Elaine Liston at the KSC Archives; Dr. Roger D. Launius at the National Air and Space Museum; Robert G. Hoey; John L. Goodman; Nick Veronico; and Dave Arnold, Vicki Libis, Molly Koecher, Mike Machat, Barbara Harold, Connie Nordrum, and Monica Bahr at Specialty Press.

On the Cover: *Panoramic view of the lunar surface stitched together from multiple 70mm Hasselblad images taken by Neil Armstrong. Buzz Aldrin is removing the Laser Ranging Retro-Reflector (LRRR) from the Scientific Equipment (SEQ) Bay on the Lunar Module.* (NASA photo composite by Ron Stephano)

On the Back Cover (left): *The 363-foot tall Saturn 506 launch vehicle lifted-off from Kennedy Space Center (KSC) Launch Complex 39A at 13:32:00 (UTC), 16 July 1969.* (NASA)

On the Back Cover (right): *The three Apollo 11 crewmen await pickup by a helicopter from the USS* Hornet. *The fourth man in the life raft is a U.S. Navy underwater demolition team swimmer. All four men are wearing biological isolation garments. Apollo 11 splashed down at 16:50:35, 24 July 1969, about 934 miles southwest of Hawaii.* (NASA)

On the Back Cover: *The Apollo 11 crew patch. See page 17.*

On the Title Page: *The Apollo 11 flag was later blown over during takeoff by the LM ascent engine's exhaust plume.* (NASA)

On the Contents Page: *The seldom-seen dark side of the Lunar Module, looking into the Sun.* (NASA photo composite by Ron Stephano)

Distributed in the UK and Europe by
Crécy Publishing Ltd
1a Ringway Trading Estate
Shadowmoss Road
Manchester M22 5LH England
Tel: 44 161 499 0024
Fax: 44 161 499 0298
www.crecy.co.uk
enquiries@crecy.co.uk

CONTENTS

INTRODUCTION

In June 1964, Eugene N. Cameron and his NASA mineralogy and petrology planning team released a report that included an analogy between the upcoming Apollo lunar voyage and another historic event:

"Some time before the year 1492, a group of workmen were standing in a shipyard looking at a half-constructed craft. One of them said, 'It won't float;' another said, 'If the sea monsters don't get it first, it will fall off the edge;' while a third, more reflective than the others, said, 'What do they want to go for, anyway?'

"The Apollo Project is primarily a glorious adventure, in which Man, will for the first time, tread upon the surface of another celestial body. It will be a magnificent feat, and a milestone in the history of the human race. No other purpose, or justification, is necessary."

It was a magnificent achievement. Forty years have passed since Neil Armstrong and Buzz Aldrin landed on the Moon. Ten other men followed, each pair accompanied by a third astronaut that escaped much of the fame bestowed upon those that walked on the lunar surface. All of the astronauts were shadowed by several hundred-thousand engineers, technicians, and other workers that made the voyages possible.

Looking back, it was amazing, made even more so by the struggles of the current-day Constellation Program to duplicate the feat. Former NASA Chief Historian Roger Launius is fond of saying, "If we can land men on the Moon, why can't we land men on the Moon?" It is an excellent question, although far outside the scope of this book.

It is disturbing to those of us in the business of sending men into space that an increasing portion of the population does not believe the United States landed on the Moon. Why the "moon landing hoax" conspiracies have gained popularity is also beyond the scope of this book, and both of us were too young to have participated in Apollo. But each of us has talked to the astronauts that walked on the Moon, and if you ever look into the eyes of John W. Young – one of three humans to have been to the Moon twice – you will never doubt that he accomplished something unique, something that only 24 men have ever experienced.

Of course, Apollo was immensely more complicated than is depicted in this short photo retrospective of a single mission. For further reading, we recommend *Man On the Moon: The Voyages of the Apollo Astronauts* by Andrew Chaiken, *First Man: The Life of Neil A. Armstrong* by James R. Hansen, and *Carrying the Fire: An Astronaut's Journeys* by Michael Collins. Several good books have also been published by NASA about the program, the foremost being *Moonport: A History of the Apollo Launch Facilities and Operations* by Charles D. Benson and William Barnaby Faherty (NASA SP-4204, 1978), *Chariots for Apollo: A History of Manned Lunar Spacecraft* by Courtney G. Brooks, James M. Grimwood, and Lloyd S. Swenson, Jr. (NASA SP-4205, 1979) and *Stages to Saturn: A Technological History of the Apollo/Saturn Launch Vehicles* by Roger E. Bilstein (NASA SP-4206, 1980). For an excellent look at what the Soviets were doing at the time, see *Challenge to Apollo: The Soviet Union and the Space Race, 1945-1974* by Asif A. Siddiqi (NASA SP-2000-4408, 2000).

Dennis R. Jenkins
Cape Canaveral, Florida

Jorge R. Frank
Houston, Texas

Neil Alden Armstrong (b. 5 August 1930)   *Michael Collins (b. 31 October 1930)*   *Edwin Eugene Aldrin, Jr. (b. 20 January 1930)*

40th Anniversary Photographic Retrospective   5

THE APOLLO PROGRAM

The introduction to the *Apollo Spacecraft News Reference Manual*, released in 1965, summed it up well:

"It took 400 years of trial and failure, from da Vinci to the Wrights, to bring about the first flying machine, and each increment of progress thereafter became progressively more difficult.

"But nature allowed one advantage: air. The air provides lift for the airplane, oxygen for engine combustion, heating, and cooling, and the pressurized atmosphere needed to sustain life at high altitude. Take away the air and the problems of building the man-carrying flying machine mount several orders of magnitude. The craft that ventures beyond the atmosphere demands new methods of controlling flight, new types of propulsion and guidance, a new way of descending to a landing, and large supplies of air substitutes.

"Now add another difficulty: distance. All of the design and construction problems are re-compounded. The myriad tasks of long-distance flight call for a larger crew, hence a greater supply of expendables. The functions of navigation, guidance, and control become far more complex. Advanced systems of communications are needed. A superior structure is required. The environment of deep space imposes new considerations of protection for the crew and the all-important array of electronic systems. The much higher speed of entry dictates an entirely new approach to descent and landing. Everything adds up to weight and mass, increasing the need for propulsive energy. There is one constantly recurring, insistent theme: everything must be more reliable than any previous aerospace equipment, because the vehicle becomes in effect a world in miniature, operating with minimal assistance from Earth.

"Such is the scope of Apollo.

"Appropriately, the spacecraft was named for one of the busiest and the most versatile of the Greek gods. Apollo was the god of light and the twin brother of Artemis, the goddess of the Moon. He was the god of Music and the father of Orpheus. At his temple in Delphi, he was the god of prophecy. Finally, he was also known as the god of poetry, of healing, and of pastoral pursuits."

The Apollo emblem was a disc circumscribed by a band displaying the words Apollo and NASA. The center disc bore a large letter "A" with the constellation Orion positioned so its three central stars form the bar of the letter. To the right was a sphere of the Earth, with a sphere of the Moon at the upper left. The face on the Moon represented the mythical god, Apollo. A double trajectory passes behind both spheres and the central stars. (NASA)

Although the third American-manned space program to fly, Apollo was first proposed in early 1960, during the Eisenhower administration, to follow Mercury; Gemini was conceived later. While the Mercury capsule could only support one astronaut on a limited Earth orbital mission, the Apollo spacecraft was intended to carry three astronauts on a circumlunar flight and perhaps, eventually, to a lunar landing.

While NASA went ahead with planning for Apollo, funding for the program was far from certain, particularly given President Dwight D. Eisenhower's tepid response toward manned spaceflight. When John F. Kennedy was elected President in November 1960, he promised technical superiority over the Soviet Union, particularly in long-range missiles, missile defense, and space exploration. However, despite his campaign rhetoric, Kennedy did not immediately embrace Apollo. The young president knew little about the technical details and was discouraged by the massive financial commitment needed for such an ambitious undertaking. When NASA Administrator James E. Webb requested a 30-percent budget supplement to increase the pace of manned spaceflight, Kennedy supported an acceleration of the large launch vehicle (Saturn) project, but deferred a decision on the broader issue of Apollo.

Things changed quickly. On 12 April 1961, cosmonaut Yuri Gagarin became the first man in space, reinforcing American fears about being left behind in a technological competition with the Soviet Union. A meeting of the U.S. House Committee on Science and Astronautics the following day saw suggestions for a concerted effort to overtake the Soviets. Kennedy, however, was more circumspect, refusing to make any commitment until he understood the implications. On 20 April, Kennedy asked Vice President Lyndon B. Johnson to look into possible space programs that could offer the opportunity to showcase American technical superiority. Johnson, who had served as the chairman of the Senate Special Committee on Space and Astronautics and was already prepared for such a request, responded the following day, "We are neither making maximum effort nor achieving results necessary if this country is to reach a position of leadership." He concluded that a manned Moon landing was far enough in the future to make it possible that the United States could achieve it before the Soviets. Webb, warning it would be extraordinarily difficult and incomprehensively expensive, agreed that it could be done.

On 25 May 1961, Kennedy presented a Special Message to the Congress on Urgent National Needs:

"Mr. Speaker, Mr. Vice President, my copartners in Government, gentlemen, and ladies:

"The Constitution imposes upon me the obligation to 'from time to time give to the Congress information of the State of the Union.' While this has traditionally been interpreted as an annual affair, this tradition has been broken in extraordinary times.

"These are extraordinary times. And we face an extraordinary challenge. Our strength as well as our convictions have imposed upon this nation the role of leader in freedom's cause. No role in history could be more difficult or more important. We stand for freedom."

[The first eight parts of the speech concerned military and security needs, and are not reproduced here.]

"Finally, if we are to win the battle that is now going on around the world between freedom and tyranny, the dramatic achievements in space which occurred in recent weeks should have made clear to us all, as did the Sputnik in 1957, the impact of this adventure on the minds of men everywhere, who are attempting to make a determination of which road they should take. Since early in my term, our efforts in space have been under review. With the advice of the Vice President, who is Chairman of the National Space Council, we have examined where we are strong and where we are not, where we may succeed and where we may not. Now it is time to take longer strides – time for a great new American enterprise – time for this nation to take a clearly leading role in space achievement, which in many ways may hold the key to our future on Earth.

"I believe we possess all the resources and talents necessary. But the facts of the matter are that we have never made the national decisions or marshalled the national resources required for such leadership. We have never specified long-range goals on an urgent time schedule, or managed our resources and our time so as to insure their fulfillment.

"Recognizing the head start obtained by the Soviets with their large rocket engines, which gives them many

President John F. Kennedy addresses a joint session of Congress, on 25 May 1961, committing the United States to landing a man on the Moon. In the background are Vice President Lyndon B. Johnson (left) and Speaker of the House Sam T. Rayburn. (NASA)

months of leadtime, and recognizing the likelihood that they will exploit this lead for some time to come in still more impressive successes, we nevertheless are required to make new efforts on our own. For while we cannot guarantee that we shall one day be first, we can guarantee that any failure to make this effort will make us last. We take an additional risk by making it in full view of the world, but as shown by the feat of astronaut Shepard, this very risk enhances our stature when we are successful. But this is not merely a race. Space is open to us now; and our eagerness to share its meaning is not governed by the efforts of others. We go into space because whatever mankind must undertake, free men must fully share.

"I therefore ask the Congress, above and beyond the increases I have earlier requested for space activities, to provide the funds which are needed to meet the following national goals:

"First, I believe that this nation should commit itself to achieving the goal, before this decade is out, of landing a man on the Moon and returning him safely to the Earth. No single space project in this period will be more impressive to mankind, or more important for the long-range exploration of space; and none will be so difficult or expensive to accomplish.

"We propose to accelerate the development of the appropriate lunar spacecraft. We propose to develop alternate liquid and solid fuel boosters, much larger than any now being developed, until certain which is superior. We propose additional funds for other engine development and for unmanned explorations – explorations which are particularly important for one purpose which this nation will never overlook: the survival of the man who first makes this daring flight. But in a very real sense, it will not be one man going to the Moon – if we make this judgment affirmatively, it will be an entire nation. For all of us must work to put him there.

"Secondly, an additional 23 million dollars, together with 7 million dollars already available, will accelerate development of the Rover nuclear rocket. This gives promise of some day providing a means for even more exciting and ambitious exploration of space, perhaps beyond the Moon, perhaps to the very end of the solar system itself.

"Third, an additional 50 million dollars will make the most of our present leadership, by accelerating the use of space satellites for world-wide communications.

"Fourth, an additional 75 million dollars – of which 53 million dollars is for the Weather Bureau – will help give us at the earliest possible time a satellite system for world-wide weather observation.

"Let it be clear – and this is a judgment which the Members of the Congress must finally make – let it be clear that I am asking the Congress and the country to accept a firm commitment to a new course of action, a course which will last for many years and carry very heavy costs: 531 million dollars in fiscal '62 – an estimated seven to nine billion dollars additional over the next five years. If we are to go only half way, or reduce our sights in the face of difficulty, in my judgment it would be better not to go at all.

"Now this is a choice which this country must make, and I am confident that under the leadership of the Space Committees of the Congress, and the Appropriating Committees, that you will consider the matter carefully.

"It is a most important decision that we make as a nation. But all of you have lived through the last four years and have seen the significance of space and the adventures in space, and no one can predict with certainty what the ultimate meaning will be of mastery of space.

"I believe we should go to the Moon. But I think every citizen of this country as well as the Members of the Congress should consider the matter carefully in making their judgment, to which we have given attention over many weeks and months, because it is a heavy burden, and there is no sense in agreeing or desiring that the United States take an affirmative position in outer space, unless we are prepared to do the work and bear the burdens to make it successful. If we are not, we should decide today and this year.

"This decision demands a major national commitment of scientific and technical manpower, materiel and facilities, and the possibility of their diversion from other important activities where they are already thinly spread. It means a degree of dedication, organization, and discipline, which have not always characterized our research and development efforts. It means we cannot afford undue work stoppages, inflated costs of material or talent, wasteful interagency rivalries, or a high turnover of key personnel.

"New objectives and new money cannot solve these problems. They could in fact, aggravate them further – unless every scientist, every engineer, every serviceman, every technician, contractor, and civil servant gives his personal pledge that this nation will move forward, with the full speed of freedom, in the exciting adventure of space."

Humble Beginnings

Kennedy's decision came after a long series of Soviet firsts. They were the first to launch a satellite, and the first to send an unmanned probe to the Moon; only a month earlier, Yuri Gagarin had become the first man in space. Alan B. Shepard followed Gagarin by only a few weeks to become the first American in space, although this was a suborbital flight and not directly comparable in technical merit. Nevertheless, Shepard's flight had given Americans the feeling that, with a major effort, they might close the gap with the Soviets. The public was ready and willing to expand the space effort, and Congress did not blink at Webb's estimate that $20 to $40 billion could be required to go to the Moon. It was a popular decision, and the vote of the Congress on the program was virtually unanimous.

In general, Kennedy felt great pressure to have the United States "catch up to and overtake" the Soviet Union in the newly defined "space race." However, Kennedy's decision was not based, and least not completely, on a desire to advance space exploration. Instead, he felt the United States needed to embark on a non-military battle to win the hearts and minds of developing nations to keep them from adopting communist governments. The Soviets had tested an atomic bomb in 1949 and their first thermonuclear weapon in 1955, eliminating

Only 20 days before Kennedy's speech, Deke Slayton (far left) and Virgil Grissom (far right) were on hand to greet Alan B. Shepard (in flight suit) on Grand Bahama Island after the first U.S. manned suborbital flight. Just behind Shepard is Dr. Keith Lyndell. (NASA)

perceived American military advantages The ill-fated Bay of Pigs excursion in mid-April 1961 had seriously damaged the reputation of Kennedy and the United States, and the goodwill from the Berlin Airlift was fading. Something was needed to show American superiority over the Soviets. Thus, the Cold War, not space exploration, is the primary context through which many historians now view Kennedy's speech.

Landing men on the Moon required the largest commitment of resources ever made by a nation in peacetime. Only the construction of the Panama Canal in 1914 and the development of the atomic bomb during World War II were comparable in scope. At its peak, the Apollo program employed 400,000 people and required the support of over 20,000 industrial firms and universities.

Once Kennedy had defined the goal, several major decisions had to be made. Perhaps the most daunting, and certainly the most important in retrospect, was the mission mode necessary to get to the Moon. This decision would largely determine the size of the required booster and the number of launches necessary for each mission. Four possibilities were considered:

Lunar Surface Rendezvous: Two spacecraft would be launched for each mission. The first would be an automated vehicle carrying propellants. After a successful landing by the supply vehicle, a manned spacecraft would be launched and land next to the supply vehicle where it would refuel for the return to Earth. Since

the ability to precisely control where a spacecraft landed was unknown (and judging by the future Apollo 11, tricky on the first attempt), this mode was never seriously considered.

Direct Ascent: A spacecraft would travel directly to the Moon, landing and returning as a unit. Initially, this was the favored mode, but it would have required an inconceivably powerful launch vehicle that Wernher von Braun did not believe could be built using foreseeable technology.

Earth Orbit Rendezvous (EOR): Two Saturn Vs would be launched, one carrying the manned spacecraft and one carrying a propulsion unit that would enable the spacecraft to escape Earth orbit. After a docking in Earth orbit, the spacecraft would land on the Moon as a unit. This was the leading contender during most early planning, explaining the vast infrastructure built at the Kennedy Space Center.

Lunar Orbit Rendezvous (LOR): One Saturn V would launch a spacecraft composed of modular parts. A command module would remain in orbit around the Moon, while a lunar excursion module would descend to the Moon and then return to dock with the command module in lunar orbit. In contrast with the other

Building the Panama Canal was the only other peacetime project similar in scope to the effort to land men on the moon. Three sets of locks connect natural rivers and lakes to allow ships to transit between the Atlantic and Pacific Oceans below the Puente de las Américas that connects North and South America. After the United States took over the canal project, it took roughly ten years to complete, approximately the same as Apollo. This is the construction site at the approach wall of the Miraflores Locks in 1913. (Library of Congress)

plans, LOR required only a small part of the spacecraft to land on the Moon, minimizing the mass to be launched from the Moon's surface for the return trip, and significantly reducing the amount of propellant needed to be carried from Earth.

Prior to the Kennedy's speech, NASA generally favored the Direct Ascent mode. Many engineers feared that the two rendezvous modes would be too difficult given that a rendezvous had never been attempted in space. However, once von Braun determined that a booster large enough to support Direct Ascent could not be built within the required time frame, Earth Orbit Rendezvous became the favored mode – engineers figured that if something went drastically wrong during rendezvous, the manned spacecraft was close enough to make an emergency return.

However, dissenters, such as John C. Houbolt at the NASA Langley Research Center, emphasized the significant weight reductions that were offered by the LOR approach. Throughout 1960 and 1961, Houbolt, and others, campaigned for the recognition of LOR as a valid and practical option. While acknowledging that he spoke "somewhat as a voice in the wilderness," Houbolt pleaded that LOR should not be discounted without serious consideration.

In the LOR mode, eliminating the propulsion system and propellants needed to soft-land the entire vehicle on the lunar surface would cut the spacecraft weight in half. Instead, a small lunar excursion module (later shortened to lunar module) would carry two of the three-man crew to a soft landing on the Moon. A portion of the lunar module would subsequently be launched from the Moon to rendezvous with the third crewmember still in lunar orbit in the command module. The entire crew would then return to Earth aboard the command module.

By early 1962, members of the NASA Space Task Group at the Manned Spacecraft Center in Houston began to support LOR. The engineers at Marshall Space Flight Center, in charge of the Saturn launch vehicles, took a little longer, but their support for LOR was announced by von Braun at a briefing in June 1962. The reasons for the change in support were many: LOR was thought to provide a higher probability of success with essentially equal safety, promised mission success months earlier than the other modes, would cost 10 to 15 percent less than the other modes, and required the least amount of technical development. The decision in favor of LOR was formally announced on 11 July 1962.

Noted space historian James Hansen concludes, "Without NASA's adoption of this stubbornly held minority opinion in 1962, the United States may still have reached the Moon, but almost certainly it would not have been accomplished by the end of the 1960s."

Mission Types

In September 1967, the Manned Spacecraft Center outlined seven Apollo mission types that would lead up to a manned lunar landing. Each mission type would test a specific set of components and tasks; each previous step needed to be completed successfully before the next mission type could be undertaken. These were:

A – Unmanned Command/Service Module (CSM) test flight
B – Unmanned Lunar Module (LM) test
C – Manned CSM in low Earth orbit
D – Manned CSM and LM in low Earth orbit
E – Manned CSM and LM in an elliptical Earth orbit with an apogee of 4,600 miles
F – Manned CSM and LM in lunar orbit
G – Manned lunar landing with a single, short extravehicular activity (EVA)

Three additional types were subsequently identified:

H – Short duration stays on the Moon with two EVAs
J – Longer three-day stays, with three EVAs and the use of a lunar rover
I – Long duration orbital missions using a Service Module bay loaded with scientific equipment

The first test flight of the Saturn V booster, Apollo 4, on 9 November 1967, exemplified George Mueller's strategy of "all up" testing. Rather than being tested stage by stage, the Saturn V was flown for the first time as one unit; the flight was successful. As were all future Apollo missions, the launch was covered live by Walter Cronkite on CBS, the *de facto* spokesman for the American space program.

The unmanned Apollo 4 and Apollo 6 were A-missions, each launching a Block I Command and Service Module (CSM) into Earth orbit. Apollo 5 was the B-mission, testing an unmanned Lunar Module (LM) in Earth orbit. Apollo 7, launched on 11 October 1968, was a manned Earth orbiting flight of the CSM, completing the objectives for the C-mission.

However, by the summer of 1968, it became clear that a fully functional LM would not be available for the first D-mission, Apollo 8. Rather than perform a repeat of the Apollo 7 Earth-orbiting mission, NASA decided to create a so-called C-Prime mission by sending Apollo 8 around the Moon during Christmas 1968. Apollo 9, launched on 3 March 1969, tested the CSM and LM in Earth orbit, fulfilling the objectives of the D-mission. The E-mission was cancelled before it was flown since it seemed to add little to the confidence level of the program and took valuable time that was not available after the delay recovering from the Apollo 1 fire. Apollo 10, the F-mission, was launched on 18 May 1969. This dress rehearsal for a Moon landing brought Thomas P. Stafford and Eugene A. Cernan in the LM to within 10 miles of the lunar surface while John W. Young orbited in the CM. The mission fulfilled all of the objectives of Kennedy's speech except for the landing. That distinction would go to the first G-mission – Apollo 11.

Saturn V Description

When the United States decided to go to the Moon, no available launch vehicle even remotely approached the needed capability. Mercury was using hastily converted military missiles – Redstone and Atlas – to launch the small, single-man capsules, and even the heavyweight Titan ICBM on the drawing board could not lift any significant mass to the Moon.

However, engineers at the Army Ballistic Missile Agency in Hunstville, Alabama, under the leadership of Wernher von Braun, had been thinking about larger launch vehicles for several years. In April 1957, the Germans had begun studying a family of launch vehicles that clustered existing rocket engines. The following year, von Braun's group began working with the Rocketdyne Division of North American Aviation to create the 1,500,000-lbf kerosene-oxygen F-1, the 200,000-lbf kerosene-oxygen H-1, and the 200,000-lbf hydrogen-oxygen J-2 rocket engines. The F-1 would become, by far, the largest and most powerful rocket engine ever developed.

On 15 August 1958, the Department of Defense Advanced Research Projects Agency (ARPA) approached the Army Ballistic

The Vehicle Assembly Building (VAB) under construction at Launch Complex 39 on Merritt Island, Florida. This 9 November 1964 view shows the four high bays that would be used to stack the Saturn V launch vehicle are still being erected. The white Launch Control Center (LCC) on the right was partially built, and the red Launch Umbilical Towers on the Mobile Launchers can be seen behind the construction. This facility currently supports the Space Shuttle Program and will soon support the Constellation Program's attempt to return to the Moon. (NASA)

Missile Agency to develop a large launch vehicle. The project was named Saturn on 3 February 1959. With the passing of the Space Act in 1957, the military was supposed to get out of the space business and the newly formed NASA assumed technical direction of Saturn in late 1959. The entire project was administratively transferred to NASA on 16 March 1960, and the Army development group at Huntsville became the nucleus of the new NASA Marshall Space Flight Center (MSFC) on 1 July 1960.

The largest of the vehicles under active development, the Saturn I (called the C-1 at the time), was still much too small to offer any real hope of sending Apollo to the Moon, except possibly by using half-a-dozen launches and assembling the spacecraft in Earth orbit. At the time of Kennedy's speech, the Saturn I was still six months away from its first test flight, although its initial static firing had taken place on 29 April 1960.

Fortunately, the Germans were also looking at larger vehicles. The C-2 design was stillborn, but the C-3 would require only four launches for the Earth Orbit Rendezvous concept and the C-4 would need only two launches. In the end, none of these would be built since NASA decided to go directly to the largest of the proposed family of boosters: the C-5.

On 10 January 1962, NASA announced plans to build the C-5 that could lift 250,000 pounds into low-earth orbit, or nearly 100,000 pounds into a translunar trajectory, easily supporting the Lunar Orbit Rendezvous concept. The C-5 was confirmed as the choice for Apollo in early 1963, and was given a new name – Saturn V. To reduce the time needed to develop the new launch vehicle, the Saturn V third stage (S- IVB) was patterned after the Saturn I second stage (S-IV) and the Saturn V instrument unit was an outgrowth of the one used on Saturn I.

The Saturn V was the first U.S. launch vehicle specifically designed for manned missions (the Saturn I was designed as general-purpose booster). Including the Apollo spacecraft, the Saturn V was 364 feet high and weighed approximately 6.1 million pounds fully loaded. The vehicle consisted of three stages and an Instrument Unit that contained the guidance system. The S-IC first stage was 33 feet in diameter and used five F-1 engines arranged in a cruciform pattern to produce 7,500,000-lbf for approximately 2.5 minutes. The stage consumed 4.5 million pounds of propellants and pushed the spacecraft to just over 5,000 mph. The S-II second stage was also 33 feet in diameter and used five J-2 engines that burned for approximately six minutes. This stage took the spacecraft to a near orbital velocity of 15,300 mph and an altitude of about 115 miles. At this point the 21.75-foot diameter S-IVB third stage took over, taking 2.75

minutes to get the spacecraft into orbit, approximately 12 minutes after liftoff. Its single J-2 engine could be restarted in orbit to make the 5-minute trans-lunar injection (TLI) burn that would send the Apollo spacecraft toward the Moon.

The instrument unit, 21.75 feet in diameter and 3 feet long, sat atop the third stage, weighed approximately 4,500 pounds, and contained the electronic equipment that controlled all three stages and their engines. The Apollo spacecraft was located directly above the instrument unit.

The first manned Apollo mission was Apollo 7, on 11 October 1968, using a Saturn IB launch vehicle instead of the much more powerful Saturn V that would be used for the Moon missions. Launch Complex 34 had been the site of the Apollo 1 fire. (NASA)

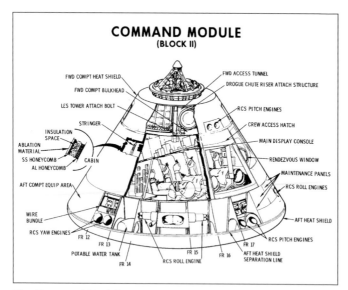

COMMAND MODULE
(BLOCK II)

FWD COMPT HEAT SHIELD
FWD COMPT BULKHEAD
LES TOWER ATTACH BOLT
STRINGER
INSULATION SPACE
ABLATION MATERIAL
SS HONEYCOMB
AL HONEYCOMB
CABIN
AFT COMPT EQUIP AREA
WIRE BUNDLE
RCS YAW ENGINES
FR 12
FR 13
POTABLE WATER TANK
FR 14
RCS ROLL ENGINE
FR 15
FR 16
AFT HEAT SHIELD SEPARATION LINE
FR 17
AFT HEAT SHIELD
RCS PITCH ENGINES
RCS ROLL ENGINES
MAINTENANCE PANELS
RENDEZVOUS WINDOW
MAIN DISPLAY CONSOLE
CREW ACCESS HATCH
RCS PITCH ENGINES
DROGUE CHUTE RISER ATTACH STRUCTURE
FWD ACCESS TUNNEL

The Command Module (CM) was remarkably small to support three men for a lunar mission. The 10.55-foot-high capsule had a maximum diameter of 12.85 feet at its base. The pressurized compartment provided approximately 210 cubic feet of volume for the crew and equipment. Three couches faced forward in the center of the compartment; with the seat portion of the center couch folded, two astronauts could stand at the same time. A hatch was located above the center couch and a short tunnel led to the docking port used to mate with the Lunar Module. (NASA)

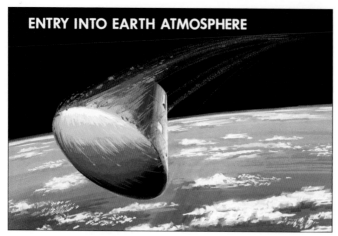

ENTRY INTO EARTH ATMOSPHERE

The lunar return trajectory of Apollo 11 resulted in an atmospheric entry velocity of 36,194 feet per second (24,677 mph), which created an aerodynamic heating environment approximately four times as severe as that experienced by Mercury and Gemini. (NASA)

The Boeing Company, the Space Division of North American Aviation, and the Douglas Aircraft Company were the prime contractors for the Saturn V first, second, and third stages, respectively. The engines for all three stages were developed and manufactured by the Rocketdyne Division of North American. The Instrument Unit was produced by International Business Machines.

Apollo Spacecraft Description

The Apollo spacecraft consisted of two major parts, the Command Module (CM) and Service Module (SM), collectively called the Command and Service Module (CSM). The CSM was designed and manufactured by the Space Division of North American Aviation. A Spacecraft-LM Adapter (SLA) sat between the S-IVB stage and the CSM and carried the Lunar Module. In addition, a Launch Escape System was located above the CM during ascent and could pull the capsule away from the Saturn V in the event of a catastrophic failure of the Saturn IB or Saturn V launch vehicle.

The original Block I Command Modules were found to contain several serious flaws, a combination of which resulted in the loss of Virgil I. "Gus" Grissom, Edwards H. White II, and Roger B. Chaffee in the AS-204 (Apollo 1) fire at Launch Complex 34 on 27 January 1964. Many aspects of the CM were redesigned after the fire, and the resulting Block II vehicles were used for all manned Apollo flights.

One of the most significant changes was the elimination of the 5-psi pure-oxygen atmosphere while the spacecraft was on the ground and during ascent, a change North American had recommended years before the fire. In Block II, the cabin atmosphere was sea-level pressure (14.7 psi) and consisted of 60-percent oxygen and 40-percent nitrogen prior to launch, lowering to 5 psi during ascent, and gradually changing to 100-percent oxygen during the first 24 hours of trans-lunar coast.

The CM housed the crew, life support systems, avionics, and reentry equipment. The SM carried most of the consumables (oxygen, water, helium, fuel cells, and propellant) as well as the Service Propulsion System (SPS) engine that provided the final push into lunar orbit, and the power to leave lunar orbit and return to Earth. Despite the entire vehicle weighing 6.1 million pounds at launch, only the Command Module was recovered at the end of the mission, with a splashdown weight of approximately 11,000 pounds.

Command Module

The CM consisted of an inner crew compartment and an outer heat shield. The inner shell was aluminum honeycomb sandwiched between aluminum alloy sheets and the outer shell was stainless steel honeycomb between stainless steel sheets. The outside of the capsule was covered with a non-receding, charring ablative heat shield. A layer of insulation separated the two shells. At the top of the cone was a hatch and docking assembly designed to mate with the lunar module. The CM had a maximum diameter of 12.8 feet at its base and was 10.55 feet high. It weighed approximately 13,000 pounds at liftoff (including the crew) and 11,000 pounds at splashdown.

The capsule used a spherical section forebody heatshield with a converging conical afterbody. The CM flew a lifting entry with a hypersonic trim angle of attack of −27° (0° is blunt-end first) to yield a reconstructed lift-over-drag (L/D) of 0.282 for Apollo 11. This angle-of-attack was achieved by precisely offsetting the vehicle's center-of-mass from its axis of symmetry. The heat shield on the blunt end consisted of brazed stainless steel honeycomb filled with a phenolic epoxy resin ablator and varied in thickness from 0.7 to 2.7 inches.

The CM was divided into three compartments: forward, crew, and aft. The forward compartment was the relatively small area at the apex of the cone, the pressurized crew compartment occupied most of the center section of the structure, and the aft compartment was another relatively small area around the periphery of the capsule near the base.

The forward compartment held the parachute recovery system. This consisted of two 16.5-foot-diameter white nylon conical-ribbon drogue parachutes that deployed at 23,000 feet to orient and slow the spacecraft to 175 mph so that the main parachutes could be safely deployed. The drogue chutes pulled out three 7.2-foot white nylon ring-slot parachutes that deployed the three 83.5-foot orange-and-white-striped ringsail main parachutes. The main chutes deployed at 10,000 feet to reduce the speed of the spacecraft from 175 mph to 22 mph when it entered the water.

The crew compartment occupied approximately 210 cubic feet, and held the controls, displays, navigation equipment, and other systems used by the astronauts. Three astronaut couches were lined up facing forward in the center of the compartment. With the seat portion of the center couch folded, two astronauts could stand at the same time. Generically, the astronaut in the left-hand couch was the

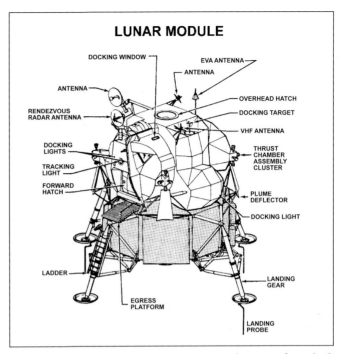

LUNAR MODULE

The Lunar Module (LM) is the only true manned spacecraft yet built since it contained none of the compromises usually necessary for flight in the Earth's atmosphere. The LM was extremely light and very fragile while it was on Earth, but functioned remarkably well in the vicinity of the Moon. (NASA)

Artist concept of the ascent stage of the Lunar Module lifting-off to return the crew to the orbiting Command Module. Hypergolic propellants ignited on contact with each other, eliminating any possible failures associated with an ignition system. (NASA)

spacecraft commander; he normally operated the flight controls. The astronaut in the center couch was the CM pilot; his principal task was guidance and navigation, and was the astronaut who remained in the CM. The astronaut in the right-hand couch was the LM pilot and his principal task was subsystems management. A large ground access hatch was located above the center couch and a short access tunnel led to the docking hatch in the nose. The CM had five windows: one in the access hatch, one next to each astronaut in the two outer seats, and two forward-facing rendezvous windows.

The aft compartment was located around the periphery of the capsule at its widest part, just forward of (above) the aft heat shield. The compartment was divided into 24 bays that contained 10 reaction control thrusters; the fuel, oxidizer, and helium tanks for the CM reaction control subsystem; water tanks; the crushable ribs of the impact attenuation system; and a number of instruments. The umbilical that carried wiring and plumbing between the CM and SM was also in the aft compartment.

Service Module

The cylindrical SM was 12.8 feet in diameter and 25 feet long. The outer skin of the SM was formed of 1-inch-thick aluminum honeycomb panels. The gap between the CM and the forward bulkhead of the SM was closed off with a fairing that was 0.5-inch thick and 22 inches high.

At the extreme rear of the SM was the 20,500-lbf Service Propulsion System (SPS) engine. The engine was 3.4 feet long, not including the 9.3-foot long radiatively cooled extension nozzle, and used a 50-50 mixture of hydrazine and unsymmetrical dimethylhydrazine (UDMH) as the fuel and nitrogen tetroxide as the oxidizer. These propellants were hypergolic, meaning they burned on contact with each other without requiring an ignition source. Attitude control was provided by identical banks of four 100-lbf reaction-control thrusters spaced 90 degrees apart around the forward part of the SM.

The interior of the SM was divided into six sectors around a central cylinder. The sectors were of three different sizes, with two sectors each of the same size. The 360 degrees around the center section was divided into two 50-degree (Sectors 1 and 4), two 60-degree (Sectors 3 and 6), and two 70-degree (Sectors 2 and 5) compartments. The sectors held three 28-volt hydrogen-oxygen fuel cells, cryogenic propellant tanks for the fuel cells, propellant tanks for the SPS engine, and various equipment and subsystems. Two helium tanks were mounted in the central cylinder. Electrical power system radiators were at the top of the cylinder and environmental control radiator panels were spaced around the bottom.

The Service Module remained attached to the Command Module until just before entry, when it was jettisoned. Afterwards, the SM burned up during its entry.

Lunar Module

The Lunar Module is the only true manned spacecraft yet developed, making no compromises for operation in the Earth's atmosphere. The vehicle, manufactured by the Grumman Aircraft Engineering Company, was composed of a descent stage that landed the vehicle on the Moon, and an ascent stage that returned the crew to lunar orbit to rendezvous with the Command Module. Fully loaded, the LM weighed just over 33,000 pounds. It is ironic that a company with the proud nickname of "Grumman Iron Works" would manufacture one of the most fragile aerospace vehicles ever built.

The descent stage comprised the lower part of the spacecraft and was an octagonal platform 13.75 feet across and 5.6 feet thick. Four landing legs with round footpads were mounted on the sides, and the leg under the access hatch had a small astronaut egress platform and ladder. The descent stage contained the landing engine, two tanks of aerozine 50 fuel, two tanks of nitrogen tetroxide oxidizer, water, oxygen, and helium

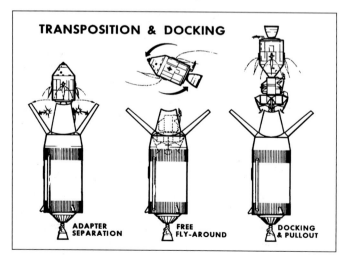

The Apollo CSM turned around to dock nose-to-top with the Lunar Module, allowing the crew to transfer between the vehicles without an extravehicular activity (EVA). (NASA)

tanks, and storage space for the lunar equipment and experiments. After servicing as a launch platform for the ascent stage, the descent stage was left on the surface of the Moon.

The ascent stage was an irregularly shaped unit approximately 9 feet high and 14 feet wide that contained the 235-cubic-foot crew compartment. There was an ingress-egress hatch in one side and a docking hatch for connecting to the CM on top. Also mounted along the top were a rendezvous radar antenna, a steerable S-band antenna, and a pair of VHF antennas. Two triangular windows were above and to either side of the egress hatch and four thruster assemblies were mounted around the sides. At the base of the assembly was the ascent engine. The stage also contained an aerozine 50 fuel tank, a nitrogen tetroxide oxidizer tank, and helium, liquid oxygen, gaseous oxygen, and reaction control system propellant tanks.

There were no seats in the LM since the two-man crew stood during descent and ascent. A control console was mounted in the front of the crew compartment above the ingress-egress hatch and between the windows. Guidance and navigation control were provided by a radar ranging system, an inertial measurement unit, and the Apollo guidance computer. The ascent stage was launched from the Moon at the end of lunar surface operations and returned the astronauts to the CSM. On most of the lunar flights, after the crew transferred to the CM, the ascent stage was sent crashing into the Moon.

Early Apollo Surface Experiments Package (EASEP)

The Early Apollo Surface Experiments Package (EASEP) was the forerunner of the Apollo Lunar Surface Experiments Package (ALSEP) used on later missions. The EASEP consisted of two solar panels that provided power (it could only operate during lunar day), a communications system to send data to Earth, a passive seismometer to measure seismic activity, and a lunar dust detector to measure dust accumulation and radiation damage to solar cells. The seismometer and dust detector were mounted on a square base, along with an isotope heater and cylindrical mast with an antenna positioning mechanism. Two brackets protruded from opposite sides of the base and held the canted rectangular solar panels, positioned to face toward east and west. The EASEP weighed approximately 105 pounds. Although not physically attached, the laser ranging retroreflector (LRRR), deployed about 16 feet north-northwest of the main experiment package, was also considered part of the EASEP.

The crew patch for Apollo 11 was designed by Mike Collins, who wanted to depict a "peaceful lunar landing by the United States." He picked an eagle as the symbol, put an olive branch in its beak, and drew a Moon background with the Earth in the distance. Initially, NASA officials thought the talons of the eagle looked too "warlike" and, after some discussion, the olive branch was moved to the claws. The crew worried that some cultures would not understand the Roman numeral XI and used the Arabic number 11 instead. They also decided not to put their names on the patch to "allow it to symbolize everyone who worked on the Moon landing."

When the Eisenhower silver dollar was issued in 1971, the patch design provided the eagle for the back of the coin; the design was retained for the Susan B. Anthony dollar in 1979.

The Apollo 11 Command Module was named Columbia, a traditional name for the United States used in song and poetry. Reportedly, it was also a reference to the "columbiad" cannon used to launch the Moonships in the Jules Verne novel From the Earth to the Moon. The Lunar Module was named Eagle to match the insignia on the crew patch.

CREW TRAINING

Apollo-era crew assignments were the responsibility of Donald K. "Deke" Slayton, the director of Flight Crew Operations at the Manned Spacecraft Center. Each mission had a prime crew, and a backup crew who could fly the mission if the prime crew could not. Slayton devised a rotation system in which each backup crew would become prime crew three missions later, but circumstances often dictated changes to this plan. Slayton described this in great detail in his autobiography, *Deke!*, with Michael Cassutt.

Mike Collins was previously the command module pilot on Frank Borman's crew for the mission that would eventually become Apollo 8. In July 1968, Collins was grounded because of a bone spur in his neck that required surgery, and was replaced by Jim Lovell. The backup crew for that mission was Neil Armstrong, commander; Buzz Aldrin, command module pilot; and Fred Haise, lunar module pilot. Collins returned to flight status in November 1968, too late to rejoin the Apollo 8 crew, but he worked to get the flight ready, and served as a CapCom during the mission.

After the successful return of Apollo 8, Slayton decided to rotate the 8 backup crew to the prime crew of Apollo 11, but substituted Collins for Haise since Collins was now available and had lost out on the lunar-orbit mission. Aldrin, who had previously trained as a lunar module pilot, would return to that role. Slayton informed Armstrong, Aldrin, and Collins of the decision on 6 January 1969. Six months of intense training followed. The circumstances of the crew's selection resulted in some changes to the standard operating procedures. For example, during launch Aldrin flew center seat, normally reserved for the command module pilot, since he had already trained for that role during Apollo 8.

Even after the assignment of the crew, there was no certainty that Apollo 11 would be the first landing; NASA required that the D-mission (Apollo 9) and F-mission (Apollo 10) be completed successfully first. However, the crew's training program proceeded under the assumption that Apollo 11 would be the G-mission that would land on the lunar surface.

In his autobiography *Carrying the Fire*, Collins vividly described NASA's approach to astronaut training:

"Being 'in training' for a flight means, to the average layman, something vaguely like being in training for a prize fight. True, there are physical aspects to space flight, […] [but] all this was small potatoes – the real game was the mental one, and it was played in the simulators. Here the battle was lost or won; here the crew 'flew' the critical phases of the flight over and over again until it seemed that every possible mistake had been made, and apologized for, and corrected, and that no further surprises in space were possible. This was the very heart and soul of the NASA system; this was where we spent our time above all other choices. Running on the beach is grand; learning geology is commendable; jungle living is amusing; the centrifuge hurts; the spacecraft ground tests are useful; but one is not to fly until the simulator has told him he is ready. No easy process this, for flying a simulator is in many ways more difficult than flying the spacecraft itself. […] Mistakes must be documented, preferably in graphic form. 'What happens if?' must be asked in a hundred different ways, and answered a thousand times. Only then, when the questions become repetitive and the answers uniformly correct, is the crew ready to fly."

This emphasis on simulator training was particularly true for the early lunar landing missions, Apollos 11–14, for which fully 56 percent of the total training hours were spent in the simulators. Previous missions needed less simulator time due to lower mission complexity; subsequent missions were able to reduce simulator time due to procedural maturity and standardization, and the need to concentrate on the increasingly complex lunar surface science operations.

The Lunar Module (LM) simulators provided an accurate representation of the LM systems and the onboard displays and controls, but did not replicate the dynamic experience of an actual lunar landing. To provide this training, NASA developed a Lunar Landing Research Facility (LLRF) at Langley, and built two types of vehicles to simulate lunar landing dynamics:

the exotic-looking Lunar Landing Research Vehicle (LLRV) and the Lunar Landing Training Vehicle (LLTV).

The idea for the LLRV actually preceded the LM itself. It originated at the NASA Flight Research Center before the lunar flight mode decision had been made and Armstrong participated in its development. Completely fortuitously, once the LOR mode decision was made, the LLRV's characteristics turned out to be very close to those of the proposed LM. The follow-on LLTV reproduced the LM even more accurately.

In retrospect, all of the Apollo commanders considered the LLRV/LLTVs to be a critical part of their training. Like most good training, the vehicles were actually more difficult to fly than the real LM since the system response was somewhat choppy and they were vulnerable to winds and turbulence. The LM itself was a pleasant surprise in comparison. The LLRV/LLTV were also quite risky and three of the five built were lost in crashes, including one piloted by Armstrong. Manned Spacecraft Center Director Robert Gilruth ordered the surviving vehicles retired as soon as possible after the Apollo 17 crew completed training.

The Apollo 11 training cycle was not particularly smooth, partly because Apollo 9 and 10 had priority in the simulators until those missions flew. All the way up to the Flight Readiness Review on 12 June 1969, there was concern that the Apollo 11 crew would not be ready. Apollo Program Manager Lt. Gen. Samuel C. Philips was ready to delay Apollo 11 to the August launch window, but Armstrong gave his word that the crew would be ready for 16 July, and they were.

On 5 July, the Apollo 11 flight control team conducted their final training simulations of the lunar descent and landing phase, with the Apollo 12 crew in the simulator. On the next-to-last run, the training team inserted a simulated "program alarm" from the LM's computer. The controllers had never seen this before, and guidance officer Steve Bales called for an abort. After the simulation was over, Flight Director Gene Kranz directed his team to research all the possible alarms and determine how each should be handled. The flight control team was going to be ready for 16 July, too.

The following pages illustrate many of the training facilities and activities undertaken by the Apollo 11 crew.

On 24 February 1969, Neil Armstrong (in white slacks), Buzz Aldrin (in blue jeans), and geologist Mike McEwen went on a field trip in west Texas near Sierra Blanca and the ruins of Fort Quitman, about 80 miles southeast of El Paso, Texas. Given the limited time they would be on the lunar surface, this was the only geology trip for the Apollo 11 crew after they were assigned to the mission, although both had previously made numerous trips. Using the arroyos in the area, McEwen taught the astronauts how to sample when a variety of rocks are spread out around a large area. (NASA)

Buzz Aldrin paddles to the shore of Lake Texoma during training at the U.S. Air Force Air Defense Command Life Support School, Perrin AFB, Texas. Lake Texoma is a large reservoir on the border between Texas and Oklahoma. Aldrin dropped into water after making a parasail ascent some 400 feet above the lake. (NASA)

Neil Armstrong went through ejection seat training at Perrin AFB. This was normal for astronauts, but seemed a bit redundant given Armstrong's history as a test pilot, including flights in the North American X-15 research airplane. (NASA)

Aldrin strapped into a parachute harness during training at Perrin AFB. Like the life raft training, this was standard Air Force pilot training used to prepare pilots for possible ejection from aircraft during flight, such as when the crew used their Northrop T-38s. (NASA)

Mike Collins during centrifuge training at the Flight Acceleration Facility in Building 29 at the Manned Spacecraft Center on 14 April 1969. This training taught the astronauts to counteract the effects of a high-g entry by using breathing techniques and muscle contractions. In his 1974 autobiography, Carrying the Fire, Collins recalled, "If you breathe normally, you find you can exhale just fine, but when you try to inhale, it's impossible to reinflate your lungs, just as if steel bands were tightly encircling your chest." (NASA photo scanned by Ed Hengeveld)

The Apollo 11 crew poses in front of the Lunar Module simulator in the Flight Crew Training Building at the Kennedy Space Center on 19 June 1969. Two simulators, one located in Houston and the other at KSC, provided the training for the LM portion of the missions. (NASA)

The Apollo 11 Moon Landing

Armstrong in the Lunar Module simulator at KSC on 11 July 1969. The simulators at KSC and JSC were quickly updated with as-flown data from Apollo 10, flown a couple of weeks earlier, to provide a realistic simulation of LM handling characteristics in the vicinity of the Moon. (NASA)

Neil Armstrong stands in the footpad at the base of the ladder during training with a LM mockup in Building 9 at the Manned Spacecraft Center in Houston on 15 April 1969. (NASA)

Armstrong holding the contingency sampler that would be used to grab a quick sample of lunar material as quickly as possible after he stepped out on the surface. (NASA)

Armstrong collects the contingency sample and removes the sample bag so that he can place it in a pocket on his right thigh. He would collect about two pounds of surface material, being careful to get far enough away from the lunar module that the soil would not have been contaminated by the residue from the descent engine exhaust. (NASA)

Aldrin (left) pours surface material from a large scoop into a sample bag during training in Houston. Each of the bags was placed into one of two large, sample return containers. Even the practice suits were pressurized, and Armstrong later reported this "helped a lot to lift the weight of the backpack." (NASA)

The mission objectives for Apollo 11 included rapidly filling one of the two sample return containers with approximately 22 pounds of material to ensure an adequate amount in case the EVA had to be cut short for any reason. The second container would be filled with "carefully selected" material including core samples. (NASA)

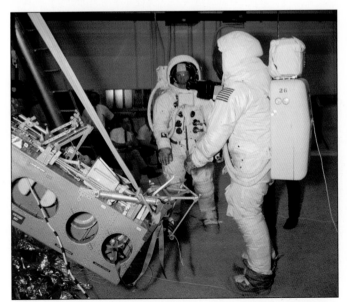

Armstrong (foreground) and Aldrin at the Modular Equipment Stowage Assembly (MESA) during training on 15 April 1969. Note the 70mm Hasselblad camera mounted on Armstrong's chest. (NASA)

Armstrong opens a sample return container on the MESA during training. Note the tools on the MESA and the large crowd of observers in the background. (NASA)

Like all of the crews, the three Apollo 11 crewmembers underwent a considerable amount of training on post-mission recovery techniques. In the photo at right, the last crewmember climbs out of an Apollo boilerplate (BP-1102) during water egress training in the Gulf of Mexico. The other two crewmen are already in the raft. The three crewmen practiced donning and wearing biological isolation garments as a part of the exercise. The rescue swimmer standing up, who assisted in the training, is also wearing a biological isolation garment. (NASA)

Mike Collins in one of the three CSM mission simulators. One simulator at MSC and two at KSC provided the major portion of the CM crew training for Apollo. The simulators were capable of accurately simulating all phases of the CM operation. (NASA)

The Apollo 11 Moon Landing

Collins inside the CSM mission simulator in Building 5 in Houston practicing procedures with the Apollo docking mechanism. Collins is at the CM's docking tunnel that provided a passageway to and from the LM following docking, and after removal of the tunnel hatches, docking probe, and drogue. Each CSM mission simulator consisted of a high-fidelity representation of the spacecraft interior, an accurate presentation of exterior visual scenes through an infinity optics display system, an operator console, and a computer complex. (NASA)

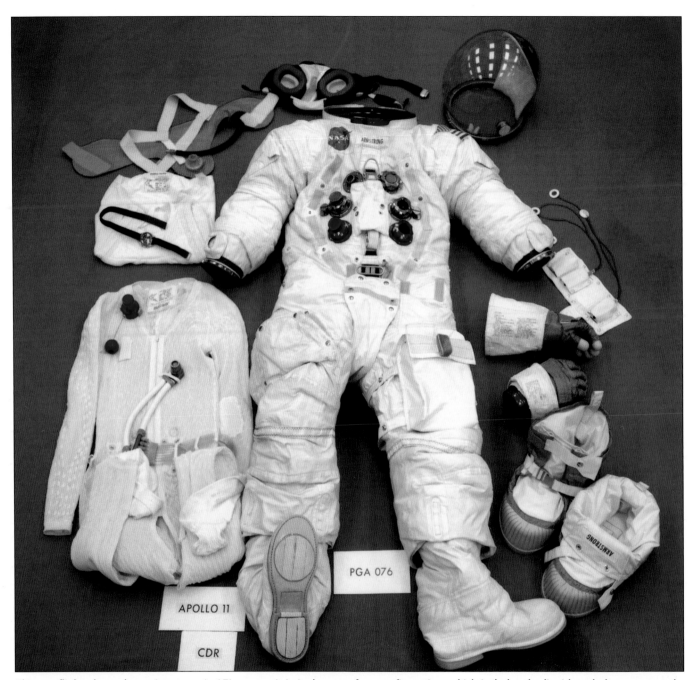

This pre-flight photo shows Armstrong's A7L spacesuit in its lunar surface configuration, which includes the liquid cooled garment at the left and the EVA gloves and boots at the right. The sewn-on cuff checklist is clearly visible on the wrist cover on the left (upper) glove. Hamilton Standard had overall development responsibility for the Apollo suit and associated portable life support system (PLSS), but the suit itself was manufactured by the International Latex Corporation (ILC). (NASA)

The Apollo 11 Moon Landing

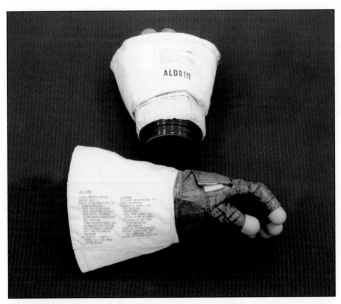

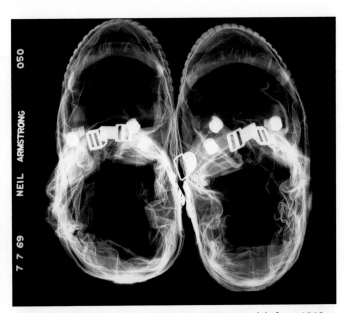

Critical checklists were sewn onto the glove cuffs, saving the astronauts from shuffling paper around in the vehicle and on the lunar surface. This pre-flight photo shows Aldrin's left lunar EVA glove. (NASA photo scanned by Eric Jones)

Jack R. Weaklandm, who worked in the MSC X-ray lab from 1968 to 1979, took this 7 July 1969 image of Armstrong's lunar EVA boots to verify that there were no foreign objects (needle points, pins etc.) embedded in the cloth. (NASA photo scanned by Ulrich Lotzmann)

The Modular Equipment Stowage Assembly (MESA) was folded against the side of the LM descent stage and was opened by Armstrong on his way down the ladder. The MESA contained all of the tools the astronauts would need on the surface. Note the large surface sample scoop on the left side of the MESA in the photo at right, and the hammer next to it (featured in the photo at left). The MESA also contained a TV camera, other tools, two sample return containers, and the surface experiments the astronauts would leave on the Moon. (NASA)

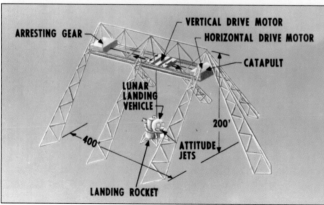

ARRESTING GEAR

VERTICAL DRIVE MOTOR

HORIZONTAL DRIVE MOTOR

CATAPULT

LUNAR LANDING VEHICLE

200'

400'

ATTITUDE JETS

LANDING ROCKET

The Lunar Landing Research Facility was located in the West Area of the Langley Research Center in Hampton, Virginia. This $3.5 million facility was constructed in 1965 to study and practice piloting problems in the final phase of the lunar landing mission. The steel A-frame structure was 400 feet long and 230 feet high and used a full-scale (but not accurate) mockup of the LM supported by an overhead partial-suspension cable system that effectively cancelled all but one-sixth of Earth's gravitational force. The remaining gravity was overcome using engines on the LM, just as would happen on the Moon. The suspension system was computer controlled so that it did not interfere with the operation of the LM. The facility is now known as the Impact Dynamics Research Facility and is used for aircraft crash tests. The site is a National Historic Landmark. (NASA)

Buzz Aldrin during training in the NASA KC-135 "Vomit Comet" aircraft. Note the LM hatch mockup in the background used to practice egress and ingress. (NASA)

Aldrin standing on the LM ladder in the KC-135 while attempting to handle a flying rock box attached to the porch rail with a lanyard. The KC-135 pilot flew precise parabolas that resulted in 1/6-g effects for about 30 seconds at a time. (NASA)

Initially, the LM used at the Lunar Landing Research Facility did not particularly resemble a real Lunar Module since it used the cockpit assembly from a Bell 47 helicopter. Like the original LLRV, this presented a problem since the pilots sat down instead of standing up as in the real LM. Hydrogen peroxide was used by the main lift engine under the center of the vehicle and the 20 attitude thrusters located around the periphery of the vehicle frame. (NASA)

The LM simulator at the Langley Lunar Landing Research Facility was eventually updated to provide a better simulation of the real Lunar Module, including a stand-up piloting position. This is Neil Armstrong standing in front of the revised vehicle on 12 February 1969. (NASA)

The Apollo 11 Moon Landing

To supplement the effort undertaken at the Langley Lunar Landing Research Facility, Bell Aerosystems built two Lunar Landing Research Vehicles (LLRV) and three Lunar Landing Training Vehicles (LLTV). This is the first LLRV on 20 August 1964 at Edwards AFB, California. (NASA)

The first LLRV in flight above Edwards on 19 August 1966 with Joe Algranti from the MSC at the controls. The General Electric CF700 lift engine – seen pointing straight down – provided 5/6ths of the lift needed for flight, effectively simulating lunar gravity. (NASA)

Initially, the LLRVs had mostly open cockpits that afforded good visibility but was not representative of the LM. Late in its life, the second LLRV was modified to partially enclose the cockpit, but the ejection seat and sitting position were, fortunately as later events proved, retained for safety reasons. (NASA)

The Lunar Landing Training Vehicles (LLTV) generally resembled the earlier LLRVs, but had enclosed partially enclosed cockpits and updated systems. The is the second LLTV at Ellington Field in Houston on 9 February 1967. Interestingly, although he was the LM pilot, Buzz Aldrin never flew the LLRV or LLTV. (NASA)

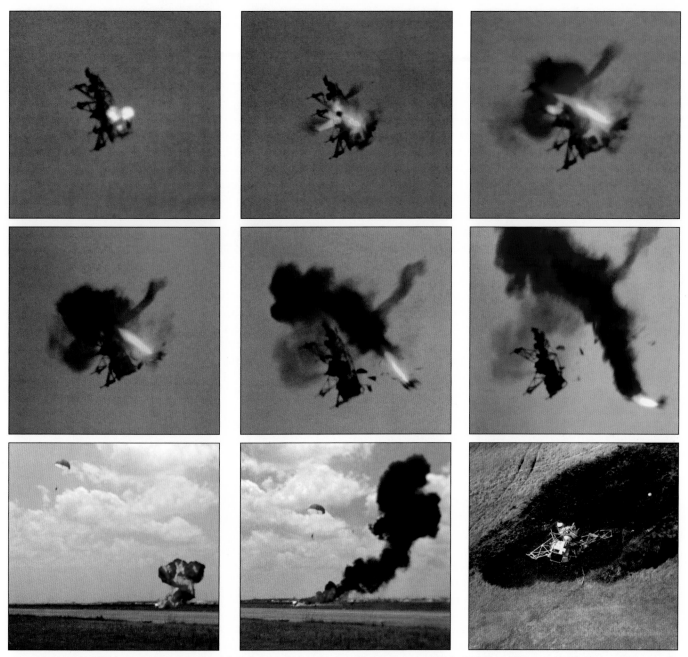

On 6 May 1968, Neil Armstrong ejected safely from the first Lunar Landing Research Vehicle (LLRV) at Ellington Field. The accident resulted, in part, from a sensor that failed to detect and warn the pilot of a fuel imbalance, which was exacerbated by gusting winds that caused the LLRV to list beyond its known recovery capability. Armstrong emerged uninjured, but the LLRV was written-off. The vehicle had made 282 flights; 198 at the NASA Flight Research Center at Edwards, California, and 84 at Ellington Field before it was lost. Armstrong had made 21 of these flights and also logged 6 flights in the second Lunar Landing Training Vehicle (LLTV). (NASA)

TO THE MOON

Apollo 11 was the 31st manned spaceflight by any nation, the 19th orbital spaceflight by the United States, the 5th manned flight of the Apollo program, the 3rd human voyage to the Moon, and the 1st manned vehicle to land on the lunar surface. The crew consisted of Spacecraft Commander Neil A. Armstrong, CM Pilot Michael Collins, and LM Pilot Edwin E. "Buzz" Aldrin, Jr. This would be the second space flight for each of the primary crew. The backup crew included James A. Lovell, William A. Anders, and Fred W. Haise.

The vehicle included the sixth Saturn V (SA-506), the seventh Block II Apollo spacecraft (CSM-107), and the fifth Lunar Module (LM-5). All of the components were stacked on Mobile Launcher 1 (ML-1) in High Bay 1 (HB-1) of the Vehicle Assembly Building at the Kennedy Space Center on Merritt Island, Florida. The splashdown on 26 May 1969, of Apollo 10 cleared the way for the first formal attempt at a manned lunar landing. The complete stack was rolled-out, at the break-neck speed of 0.9 mph, to Launch Complex 39A on 20 May 1969. The launch window, established by lighting conditions at the landing site on Mare Tranquillitatis, opened at 13:32 UTC (9:32 am, EDT) on 16 July 1969. (All times are given in Coordinated Universal Time, UTC using its French acronym, that replaced Greenwich Mean Time on 1 January 1972 as the global reference time.)

Unexpected Complication

Two days before the scheduled launch, a major problem developed. Early in the planning of Apollo, NASA decided to combine all communications between the spacecraft and Earth into a single multiplexed Unified S-Band System. This feed included all audio communications, television images, crew medical telemetry, and spacecraft telemetry. The S-Band signal was picked up by purpose-built stations at Goldstone, California, Honeysuckle Creek near Canberra, Australia, and Fresnedillas, Spain. The signals were routed to the Goddard Space Flight Center in Greenbelt, Maryland, then relayed to the Mission Control Center in Houston and other facilities.

By the late 1960s, Intelsat satellites were taking over trans-oceanic communications, and NASA ended its contracts for submarine telephone circuits, which were then reallocated by the telephone companies for normal voice use. Unfortunately, on 14 July, the INTELSAT III satellite over the Atlantic failed, cutting off the link between Spain and Goddard. Without communications, the launch of Apollo 11 would be scrubbed. Intelsat reactivated INTELSAT I (called Early Bird, since it was the first commercial communications satellite), but there were concerns the satellite might not be sufficiently reliable (it had exceeded its useful life and been deactivated in January 1969, although it remained in orbit and functional).

In an unusual instance of cooperation between governments and commercial companies, twelve transatlantic submarine telephone circuits were made available to NASA. An official with the Spanish communications authority helped the team secure the circuits with his own personal list of contacts. The last circuit was accepted by NASA only two hours before launch.

The Voyage Begins

Apollo 11 was launched at 13:32:00 UTC, 16 July 1969, and the Saturn V took several seconds to clear the Launch Umbilical Tower amid an earth-shaking rumbling that could be felt for miles. In addition to one million people crowding the highways and beaches around Central Florida, an estimated 700 million people viewed the event on television. President Richard M. Nixon viewed the proceedings from the Oval Office of the White House. As always, Walter Cronkite covered the launch live, in living color, on CBS.

The spacecraft entered Earth orbit 12 minutes after launch. After one-and-a-half orbits, the S-IVB third-stage engine ignited at 16:16:16 for a trans-lunar injection burn of 5 minutes, 48 seconds putting the spacecraft on course for the Moon. Thirty-three minutes later, the CSM separated from the third stage, turned 180 degrees, and at 16:56:03, docked with the LM that was still inside the SLA on top of the S-IVB stage. The CM pulled the LM out off the SLA, and turned around to the proper orientation for trans-lunar coast. About 75 minutes later, the S-IVB stage was injected into heliocentric orbit (a heliocentric orbit is an orbit around the Sun) where it remains.

CSM-107 is unloaded from a Super Guppy at the Cape Canaveral Skid Strip on the night of 22 January 1969. The next day it was unpacked in the Manned Spacecraft Operations Building (below). (NASA)

The descent module for LM-5 is unloaded from a Super Guppy at the Skid Strip on 17 January 1969. It took three trips to deliver a Lunar Module, one for each stage and one for the landing gear. (NASA)

The CSM for Apollo 11 after being uncrated in the Manned Spacecraft Operations Building on 24 January 1969 (NASA)

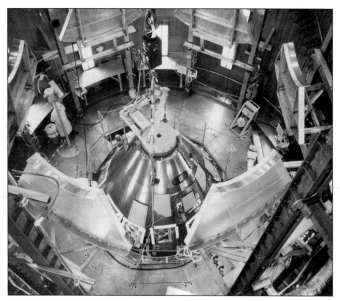

Each Apollo spacecraft was tested in one of two altitude chambers in the Manned Spacecraft Operations Building at KSC (now called the Operations and Checkout Building). This is CSM-107 for Apollo 11 being lowered into the East (left-hand) chamber. Tests with the prime crew onboard were completed on 18 March 1969, and the entire altitude chamber test series was completed on 25 March. (NASA)

In the background, CSM-107 is lifted out of the altitude chamber while CM-108 for Apollo 12 sits in the foreground. CM-107 and SM-107 had been delivered to KSC on 23 January 1969 and were mated on 29 January in preparation for the chamber tests. (NASA)

The Apollo 11 spacecraft (CSM-107) being moved into the altitude chamber at KSC. Originally, each spacecraft was to be tested in a thermal-vacuum chamber at MSC before delivery to the launch site. Because this was deemed impractical, a compromise was reached where less elaborate facilities were constructed at KSC, enabling checkout of spacecraft under modest vacuum conditions at ambient temperature as a part of normal launch preparation. These tests were adequate to confirm the operational status of all essential systems but did not involve the full thermal-vacuum environment, which might degrade thermal coatings and life-limited components to the point where refurbishment would be necessary before launch. Completed on 23 July 1965, the two chambers at KSC have an overall height of 58 feet and a diameter of 34 feet, with a 28 feet clear working area inside the chamber. The chambers are capable of reaching a test altitude of 250,000 feet. (NASA)

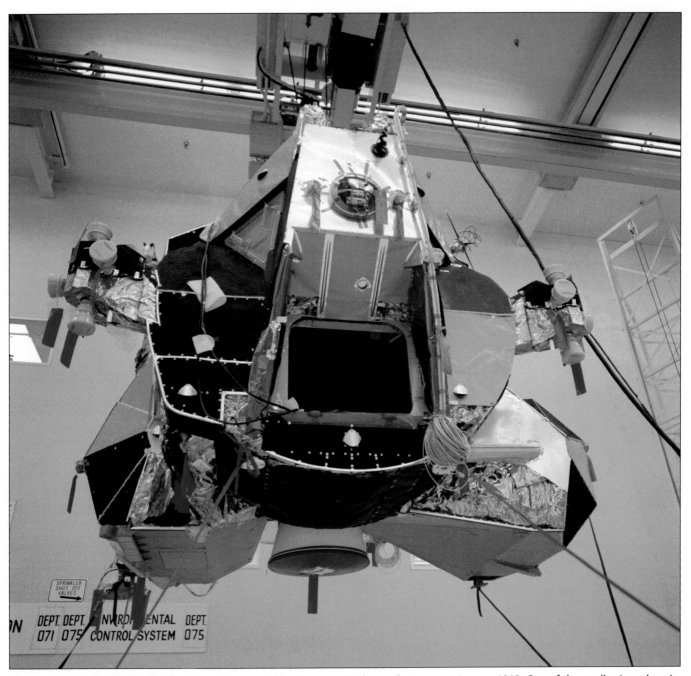

The ascent stage for LM-5 in the final assembly area at the Grumman Bethpage factory on 4 January 1969. One of the small, triangular windows may be seen at left, covered with a metal shield to protect it during manufacture. This window failed during its initial cabin pressurization test to 5.65 psi. Both inner and outer windows and the plexiglass cover of the right-hand window shattered when the pressure reached 5.1 psi. The window was replaced and the test repeated, this time successfully. (NASA)

The two LM-5 stages have been mated, but the landing gear and descent engine nozzle have not yet been installed. One of the altitude chambers may be seen in the background at left. Like the CSM, each LM was tested in an altitude chamber at KSC. Afterwards, the LM ascent stage and CSM were mated and tested again in an altitude chamber to ensure the docking mechanism sealed correctly. LM-5 altitude chamber tests were completed with the prime crew on 21 March 1969 and with the backup crew on 25 March. (NASA)

The Apollo 11 Moon Landing

LM-5, with its landing legs folded, is removed from the Landing Gear Deployment Test Facility in the Manned Spacecraft Operations Building in preparation for mating with the Spacecraft Launch Adapter. At this point, the LM still has a landing probe attached to the plus-Z footpad (at left rear, under the ladder), folded upward along the landing leg. The 67-inch-long probes were extended downward in space and told the crew when the LM touched the lunar surface. Just before the LM was installed in the Spacecraft Launch Adapter, the crew asked that the plus-Z probe be removed to ensure it did not interfere with their activities if it was bent during landing. (NASA)

Right: *Having completed its altitude chamber tests, CSM-107 is prepared for mating to Spacecraft Launch Adapter (SLA-14). The LM has already been installed in the SLA prior to the CSM being mated. The SPS engine nozzle has been added and the S-Band antenna can be seen folded downward under the SM at right. The mated spacecraft would be moved from the Manned Spacecraft Operations Building to the Vehicle Assembly Building on 14 April 1969.* (NASA)

The Boeing-built S-IC first stage of the Saturn V was 33 feet in diameter and 138 feet tall. Fully loaded with RP-1 (kerosene) and liquid oxygen (LO2), it weighed 5,030,000 pounds and its five Rocketdyne F-1 engines provided 7,500,000 pounds of thrust (lbf). The S-IC-6 stage was delivered by barge to KSC on 20 February 1969 and erected onto the Mobile Launcher the following day. (NASA)

The Apollo 11 Moon Landing

STACKING

The Vehicle Assembly Building (VAB) was created specifically to assemble, or "stack," the Saturn V launch vehicle and Apollo spacecraft. When it was built, the VAB was the largest enclosed space in the world. The building consists of a "transfer aisle" surrounded by four high-bays. Parts of the vehicles were brought into the transfer aisle horizontally, then lifted by large cranes to a vertical position. A Mobile Launcher (ML) equipped with a Launch Umbilical Tower (LUT) was placed in a high bay, and the S-IC first stage was lifted onto it. The S-II second stage and S-IVB third stage followed. Next came the Instrument Unit that contained the electronics that controlled and guided the Saturn V. The Apollo spacecraft was assembled in the Manned Spacecraft Operations Building, then brought to the VAB and lifted on top of the Instrument Unit.

The North American-built S-II second stage was 33 feet in diameter and 82 feet tall. Loaded with liquid hydrogen (LH2) and LO2, it weighed 1,060,000 pounds. Five J2 engines provided 1,150,000-lbf. (NASA)

The Douglas-built S-IVB third stage was 21.7 feet in diameter and 58.4 feet high. Loaded with LH2 and LO2, it weighed 253,000 pounds. A single Rocketdyne J-2 provided 225,000-lbf. (NASA)

Another view of the S-IVB stage being mated. The S-II-6 was erected on 4 March 1969, and the S-IVB-6 on 5 March. (NASA)

The IBM-manufactured Instrument Unit (IU) was 21.7 feet in diameter, 3 feet tall, and weighed 4,400 pounds. (NASA)

The Apollo 11 spacecraft is hoisted for mating to the SA-506 launch vehicle on 14 April. Note the LM legs. (NASA)

The Apollo 11 spacecraft is mated to the Instrument Unit. The red structure behind the spacecraft is the Launch Umbilical Tower that contained all of the umbilical connections to the launch vehicle. Note the large "white room" just to the right of the spacecraft. (NASA)

On 20 May 1969, SA-506 emerged from the Vehicle Assembly Building and started the 3.5-mile trip to Launch Complex 39A at the break-neck speed of 0.9-mph. A large, yellow hammerhead crane can be seen on top of the Launch Umbilical Tower. (NASA)

A multi-track Crawler-Transporter was used to carry the vehicle from the VAB to the launch pad. Although slow, the Crawler could carry up to 12 million pounds of Mobile Launcher, Launch Umbilical Tower, and flight vehicle, always keeping them precisely vertical. (NASA)

Personnel atop the 402-foot-high Mobile Service Structure (MSS, in the background) look back at the Apollo 11 spacecraft as the MSS is moved away during a countdown demonstration on 11 July 1969. The object on top of the capsule is the launch abort system. (NASA)

The Apollo 11 Moon Landing

The success of the Apollo 10 mission, where Thomas P. Stafford and Eugene A. Cernan had taken a Lunar Module within 10 miles of the surface, was critical to approving Apollo 11 for flight. Here the crew of Apollo 10 provides details of their experience to the Apollo 11 crew during a debriefing on 3 June 1969. (NASA)

The prime crew participate in a pre-flight press conference at KSC on 5 July 1969. From left are Armstrong, Aldrin, and Collins. The box-like enclosure surrounding the astronauts was part of elaborate precautions to reduce the possibility of exposing the crewmen to infectious disease prior to the mission. (NASA)

The Apollo 11 crew walk past the base of the Saturn V first stage during a walk-through emergency egress test on 10 June 1969. Note that each stabilizing fin on the S-IC stage has a large letter identifying it for photographic analysis. (NASA)

Armstrong and Aldrin in the LM simulator at KSC on 11 July 1969. The crew continued to use the simulators at KSC until a couple of days before launch, mostly working on contingency scenarios and evaluating the handling characteristics experienced by Apollo 10. (NASA)

A view looking down the SA-506 stack on 7 July 1969. Each of the red swing-arm umbilicals carried a specific commodity – nitrogen (N2) purge gas, liquid hydrogen (LH2), liquid oxygen (LO2), etc. – to the launch vehicle. When the first stage engines ignited and the vehicle began to move, the umbilicals quickly disconnected and swung to the side, allowing the launch vehicle to pass. The Mobile Service Structure is on the other side of the vehicle, and would be moved a safe distance away prior to launch. (NASA)

As originally conceived, all vehicle access was supposed to be accomplished in the VAB or via the Launch Umbilical Tower (LUT, at left) permanently affixed to each Mobile Launcher. However, the final configuration of the spacecraft umbilicals made it impossible to load hypergolic propellants into the CSM and LM from the LUT, leading to the development of the Mobile Service Structure (MSS) that was carried to each pad by the Crawler-Transporters. The MSS was removed from the pad area prior to launch. (NASA)

From left, William A. Anders (Apollo 8, and Apollo 11 backup CM Pilot), Armstrong, Collins, Aldrin, and Donald K. "Deke" Slayton (Director of Flight Crew Operations) during the pre-launch breakfast on 16 July 1969. Slayton is discussing last-minute details using a map – so much for a peaceful breakfast. (NASA)

The crew walking down a corridor in the Manned Spacecraft Operations Building. Note that each astronaut carried his own portable oxygen ventilator to keep his suit cool until he was connected to the environmental system in the capsule. The gloves worn inside the spacecraft were black, while the EVA gloves were white. (NASA)

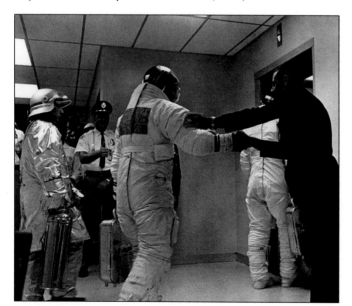

Bill Anders shakes hands with Aldrin as the crew enters an elevator in the MSOB on launch day. Note the LM tether attachment on Aldrin's right hip, a clear indication that he is an LM crewman. (NASA photo scanned by Ed Hengeveld)

Neil Armstrong leads Mike Collins and Buzz Aldrin from the Manned Spacecraft Operations Building to the transfer van for the eight-mile trip to Launch Complex 39A. Note the mission insignia on the transfer van door. (NASA)

The 363-foot-tall SA-506 begins its trek into space. The first engine (#5, in the center) ignited at 13:31:53 and all five engines were at full thrust by 13:31:58. The vehicle lifted-off at 13:32:00 and began an 8-second yaw manoeuver to ensure it cleared the Launch Umbilical Tower. The stack went supersonic 1 minute and 6 seconds after liftoff. The S-IC first stage boosted the vehicle to an altitude of 41.8 miles and 58 miles downrange in 2 minutes and 40.8 seconds. The S-IC stage separated 2 minutes and 41.6 seconds after liftoff. (NASA)

The umbilicals have swung to the side as the SA-506 launch vehicle accelerates after lift-off. Note the protective white shield covering the capsule; this would be jettisoned along with the launch escape system tower 3 minutes and 17 seconds after liftoff. (NASA)

A fish-eye lens view of the launch taken by a camera mounted on the Launch Umbilical Tower. The warm Florida humidity quickly condensed on the sides of the liquid oxygen tanks of the S-IC stage (and all the tanks on the S-II and S-IVB stage). (NASA)

A 70mm Airborne Lightweight Optical Tracking System (ALOTS) camera on an Air Force EC-135N flying at about 40,000 feet altitude captured the SA-506 vehicle as it began its pitch maneuver.

Two minutes later, the EC-135N photographed the S-IC first stage separating and the S-II second stage igniting. The discarded first stage fell into the Atlantic ocean. (U.S. Air Force)

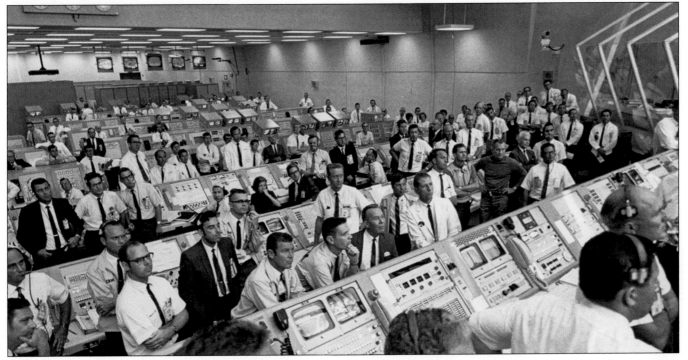

Members of the launch control team view the liftoff from Firing Room 1 in the Launch Control Center (LCC) at KSC. For Apollo, each of the control consoles faced the back of the firing room; this was reversed for Space Shuttle, where the consoles face the front, but still afford little view out of the windows that are much higher than the control room floor. Note that all of the engineers are wearing ties, and many are wearing jackets, despite it being July in central Florida. (NASA)

The Earth, as seen from Apollo 11. (NASA)

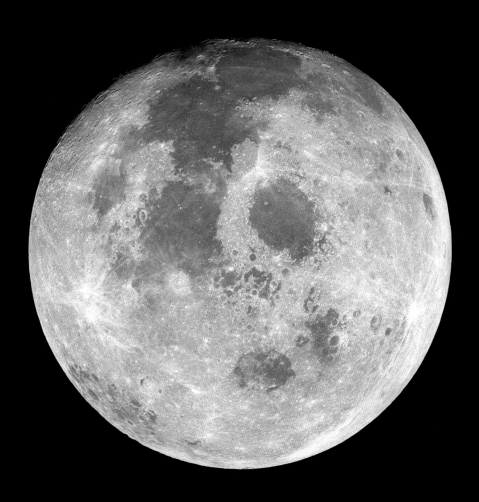

The Moon, as seen from Apollo 11. (NASA)

THE LUNAR SURFACE

During the trans-lunar coast, the crew made three color TV transmissions, and on 17 July, the SPS fired for a 3-second mid-course correction burn. A 357.5-second retrograde firing of the SPS, accomplished while the spacecraft was behind the Moon and out of contact with Earth, put the spacecraft into lunar orbit at 17:21:50 UTC on 19 July. The initial orbit was 69 by 195 miles, but two revolutions later, a 17-second SPS burn circularized the orbit at 75 by 62 miles. While in lunar orbit, the crew saw passing views of their anticipated landing site in the southern Sea of Tranquility about 12 miles southwest of the Sabine D crater. The site was selected in part because the unmanned Ranger 8 and Surveyor 5 landers, along with the Lunar Orbiter mapping spacecraft, had characterized it as relatively flat and smooth.

Armstrong and Aldrin entered the LM for final checkout, and the LM separated from the CSM at 18:11:53 on 20 July. After a visual inspection of *Eagle* by Collins in the CM, at 19:08 the LM descent engine fired for 30 seconds, putting the LM into a descent orbit with a closest approach 9 miles above the lunar surface. It had been approximately 101 hours since launch. At 20:05, the descent engine fired for 756.3 seconds, and the LM headed toward the lunar surface.

As the landing began, Armstrong reported they were "running long;" *Eagle* was 4 seconds further along its descent trajectory than planned and would land considerably west of the intended site. At an altitude of 6,000 feet, the LM guidance computer began reporting "program alarms" as it guided the descent, drawing the crew's attention from the scene outside. Inside the Mission Control Center, computer engineer Jack Garman told guidance officer Steve Bales it was safe to continue the descent in spite of the alarms.

The crew remembered the incident vividly. From a later interview with Mike Collins, "At five minutes into the burn, when I am nearly directly overhead, *Eagle* voices its first concern. 'Program Alarm,' barks Neil, 'It's a 1202.' What the hell is that? I don't have the alarm numbers memorized for my own computer, much less for the LM's. I jerk out my own checklist and start thumbing through it, but before I can find 1202, Houston says, 'Roger, we're GO on that alarm.' No problem, in other words.

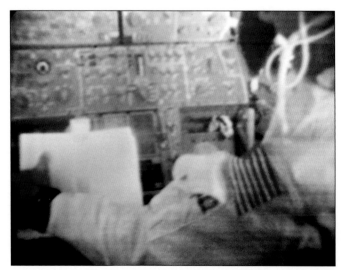 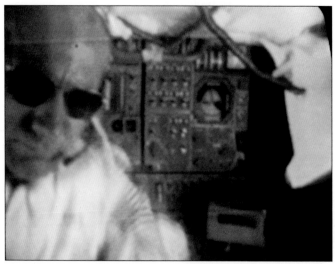

Buzz Aldrin in the LM from the third television transmission from Apollo 11 during its trans-lunar journey toward the moon. A LM window may be seen in the background of the photo at right. The LM was attached to the CSM and the docking tunnel was open during this television transmission. Apollo 11 was approximately 202,500 miles from Earth, traveling toward the moon at a speed of about 2,200 mph. (NASA)

My checklist says 1202 is an 'executive overflow,' meaning simply that the computer has been called upon to do too many things at once and is forced to postpone some of them. A little farther along, at just three thousand feet above the surface, the computer flashes 1201, another overflow condition, and again the ground is superquick to respond with reassurances."

Buzz Aldrin: "Back in Houston, not to mention on board the *Eagle*, hearts shot up into throats while we waited to learn what would happen. We had received two of the caution lights when Steve Bales, the flight controller responsible for LM computer activity, told us to proceed, through Charlie Duke, the capsule communicator. We received three or four more warnings but kept on going. When Mike, Neil, and I were presented with Medals of Freedom by President Nixon, Steve also received one. He certainly deserved it, because without him we might not have landed."

When Neil Armstrong returned his attention to the view outside, it was apparent the computer was guiding them toward a large crater with rocks scattered around it. Armstrong, "This is the area we decided we would not go into; we extended the range downrange. The exhaust dust was kicked up by the engine and this caused some concern in that it degraded our ability to determine not only our altitude in the final phases but also our translational velocities over the ground. It's quite important not to stub your toe during the final phases of touchdown."

Armstrong took control of the LM and Aldrin called out data from the radar and computer, "540 feet, down at 30 [feet per second] … down at 15 … 400 feet down at 9 … forward … 350 feet, down at 4 … 300 feet, down 3-1/2 … 47 forward … 1-1/2 down … 13 forward … 11 forward … coming down nicely … 200 feet, 4-1/2 down … 5-1/2 down … 5 percent … 75 feet … 6 forward … lights on … down 2-1/2 … 40 feet down, 2-1/2, kicking up some dust … 30 feet, 2-1/2 down … faint shadow … 4 forward … 4 forward … drifting to right a little … okay … contact light."

Eagle landed, safely, at 20:17:40 UTC (4:17:40 pm EDT) on 20 July 1969 in the region known as Mare Tranquilitatis (Sea of Tranquility). At touchdown, the LM was traveling approximately 2.1 feet per second (fps) to the crew's left and 1.7 fps vertically. The actual landing site was at 0 degree, 41 minutes, 15 seconds north latitude and 23 degrees 26 minutes and 0 seconds east longitude, compared with the planned landing at 0 degree, 3 minutes, 53 seconds north and 23 degrees, 38 minutes, and 51 seconds east. These coordinates were references to Lunar Map ORB-II-6, first edition, dated December 1967.

Aldrin began the post-landing checklist, "Contact light; Okay, engine stop; ACA – out of detent; Mode control – both auto; Override – off; Engine arm – off; 413 is in." CapCom Charles Duke interrupted, "We copy you down, *Eagle*."

Armstrong, "Houston, Tranquility Base here. The *Eagle* has landed." An emotion-filled CapCom, "Roger Tranquility. We copy you on the ground. You got a bunch of guys about to turn blue. We're breathing again. Thanks a lot."

The cause of the computer alarms was later traced to a stream of interrupts from the rendezvous radar. Although unneeded for the landing, the radar was intentionally turned on in the event of an abort. According to the Apollo 11 Mission Report (MSC-00171), post-flight analysis revealed that there was 45 seconds of fuel remaining at lunar touchdown, not as little as 7 seconds as indicated by some sources.

Armstrong remembers, "Once [we] settled on the surface, the dust settled immediately and we had an excellent view of the area surrounding the LM. We saw a cratered surface, pockmarked with craters up to 15, 20, 30 feet, and many smaller craters down to a diameter of 1 foot and, of course, the surface was very fine-grained. There were a surprising number of rocks of all sizes."

Shortly after landing, Aldrin announced, "This is the LM pilot. I'd like to take this opportunity to ask every person listening in, whoever and wherever they may be, to pause for a moment and contemplate the events of the past few hours and to give thanks in his or her own way."

It took two hours to complete the post-landing checklist, then the astronauts looked out the windows to determine where they would place the Early Apollo Scientific Experiment Package (EASEP) and the American flag. A crew rest period was planned to precede the lunar extravehicular activity (EVA), but the crew elected to perform the EVA prior to the sleep period because they were not overly tired and were adjusting easily to lunar gravity. Given the circumstance, who could sleep?

After the crew donned their portable life support systems (PLSS) and completed the required checks, Armstrong, guided through the hatch by Aldrin, began descending to the lunar surface. The controls on the chest of his suit prevented Armstrong from seeing his feet, complicating the descent. While climbing down the nine-rung ladder, Armstrong pulled a D-ring to deploy the equipment package folded against the side of the LM and activated a TV camera inside. The 210-foot antenna at Goldstone picked up the signal, but Honeysuckle Creek provided better fidelity. Nevertheless, primary coverage

The approach to Apollo Landing Site 2 in southwestern Sea of Tranquillity as seen from the LM in lunar orbit. When this photograph was taken, the LM was still docked to the CSM. Site 2 is located just right of center at the edge of the darkness. The large crater at the lower right is Maskelyne, with Maskelyne B just above it. The Ranger 8 lunar probe crashed at the edge of darkness on the right side of the image. The crater near the port forward LM RCS shadow is Censorinus K. This view looks generally west. (NASA)

of the extravehicular activity was provided by the 85-foot antenna at Goldstone, with backup coverage provided by the 85-foot antennas at Goldstone and Honeysuckle Creek. The first images used a slowscan television system that was incompatible with commercial broadcast technology so the images were played on monitors mounted in front of conventional television cameras and rebroadcast. Despite the technical difficulties, fuzzy black-and-white images of the first lunar EVA were broadcast to some 600 million people on Earth.

Approximately six and a half hours after landing, at 02:56:15 UTC on 21 July (10:56:15 pm EDT, July 20), Armstrong stepped onto the surface of the Moon, "That's one small step for [a] man, one giant leap for mankind." After describing the surface dust as "fine and powdery ... I only go in a small fraction of an inch, but I can see the footprints of my boots," he reported that moving in the reduced gravity was "even perhaps easier than the simulations ... It's absolutely no trouble to walk around."

In addition to fulfilling Kennedy's mandate to land a man on the Moon, Apollo 11 was an engineering test of the Apollo system; therefore, as Armstrong inspected the LM for damage,

he took photos so engineers would be able to evaluate its post-landing condition. He then collected a contingency soil sample, just to have one in case the LM needed to leave in a hurry, using a bag on a stick; he folded the bag and tucked it into a pocket on his right thigh. He removed the TV camera from the LM, made a panoramic sweep, and mounted it on a tripod 40 feet away.

Nineteen minutes later, Aldrin joined Armstrong on the surface, "Beautiful. Beautiful. Magnificent desolation." The astronauts then unveiled the plaque mounted on a strut behind the ladder and read the inscription aloud, "Here men from the planet Earth first set foot on the Moon July 1969, A.D. We came in peace for all mankind."

Aldrin tested methods for moving around the lunar surface, including two-footed kangaroo hops. The backpack created a tendency to tip backwards, but neither astronaut had serious problems maintaining balance. Loping became the preferred method of movement, and the astronauts reported that they needed to plan their movements six or seven steps ahead. Aldrin remarked that moving from sunlight into *Eagle's* shadow produced no temperature change inside the suit, though the helmet was warmer in sunlight, so he felt cooler in shadow.

At this point, the American flag was raised on the lunar surface, then the astronauts talked to President Nixon:

Houston: We'd like to get both of you in the field-of-view of the camera for a minute. (Pause) Neil and Buzz, the President of the United States is in his office now and would like to say a few words to you. Over.
Armstrong: That would be an honor.
Houston: All right. Go ahead, Mr. President. This is Houston. Out.
Nixon: *Hello, Neil and Buzz. I'm talking to you by telephone from the Oval Room at the White House, and this certainly has to be the most historic telephone call ever made. I just can't tell you how proud we all are of what you (garbled). For every American, this has to be the proudest day of our lives. And for people all over the world, I am sure they, too, join with Americans in recognizing what an immense feat this is. Because of what you have done, the heavens have become a part of man's world. And as you talk to us from the Sea of Tranquility, it inspires us to redouble our efforts to bring peace and tranquility to Earth. For one priceless moment in the whole history of man, all the people on*

this Earth are truly one; one in their pride in what you have done, and one in our prayers that you will return safely to Earth. [Pause]

Armstrong: *Thank you, Mr. President. It's a great honor and privilege for us to be here representing not only the United States but men of peace of all nations, and with interests and the curiosity and with the vision for the future. It's an honor for us to be able to participate here today.*

Nixon: *And thank you very much and I look forward ... All of us look forward to seeing you on the Hornet on Thursday.*

Aldrin: *I look forward to that very much, sir.*

The astronauts then deployed the solar wind experiment and set up the EASEP approximately 55 feet south of the LM. The unit was activated by ground command at 04:40:39 UTC on 21 July and was turned off about 5 hours before local lunar sunset at 10:58:46 on 3 August. The unit was activated again on the next lunar day but failed on 27 August and was terminated.

Then Armstrong loped about 400 feet from the LM to snap photos at the rim of East Crater while Aldrin collected two core samples using the geological hammer to pound in the tubes – the only time the hammer was used during the mission. The astronauts also collected rock samples using scoops and tongs on extension handles. Forty-eight pounds of core and soil samples, along with the solar wind experiment, were then packed into sample boxes.

While Aldrin and Armstrong were on the surface, Mike Collins continued orbiting the Moon in the CSM. Despite having several uninterrupted minutes each time he passed over the landing site, Collins never saw the LM on the surface. His narrow field-of-view sextant allowed him to scan only an area of approximately one square mile on each pass, but estimates of the exact position of the LM varied by several miles.

Aldrin returned to the LM first, after 1 hour and 41 minutes on the lunar surface. With some difficulty, the astronauts lifted the two sample boxes to the LM hatch using a cable pulley device called the Lunar Equipment Conveyor. Armstrong reminded Aldrin of a bag of memorial items in his suit pocket sleeve, and Aldrin tossed the bag down, after which Armstrong placed the bag on the lunar surface. The items included a gold replica of an olive branch as a traditional symbol of peace, the Apollo 1 patch, and a silicon message disk that had messages from Presidents Eisenhower,

Kennedy, Johnson, and Nixon and messages from 73 world leaders. In his 1989 book, *Men from Earth*, Aldrin says that the items included Soviet medals commemorating Cosmonauts Vladimir Komarov and Yuri Gagarin.

Separately, Armstrong's Personal Preference Kit carried a piece of wood from the left propeller of the Wright Brothers' 1903 airplane and a piece of fabric from its wing. Armstrong also carried a diamond-studded astronaut pin originally intended to be flown on Apollo 1 and given to Deke Slayton after the mission, However, following the fire, the widows gave the pin to Slayton and Armstrong took it on Apollo 11.

About 12 minutes after Aldrin, at 05:09:32 UTC, Armstrong jumped to the third rung of the ladder and climbed into the LM. The EVA ended at 05:11:13 UTC when the LM hatch was closed. After transferring to LM life support, the astronauts lightened the ascent stage by discarding their PLSS backpacks, lunar overshoes, one Hasselblad camera, and other equipment. They then pressurized the LM.

While moving in the cabin, Aldrin accidentally broke the circuit breaker handle that armed the main engine and there was initial concern this would prevent firing the engine. Fortunately, a felt-tip pen was sufficient to activate the switch. Needless to say, the breaker was modified in future LMs.

After about seven hours of fitful rest, the astronauts were awakened by Mission Control to prepare for the return flight. Two and a half hours later, the LM lifted off at 17:54:01 UTC on 21 July after 21 hours and 36 minutes on the lunar surface. *Eagle* went straight up for about 10 seconds to clear the descent stage, and then pitched over into a initial 50-degree climb. The engine fired for 7 minutes, inserting the LM into a 48.0 by 9.4-mile orbit. The LM then began maneuvering to rendezvous with the CSM in lunar orbit. As the ascent stage approached within 100 feet of the CSM, Armstrong slowed so Collins could inspect the ascent stage before docking.

Approximately 4.5 hours after lunar module ascent, Collins in the CSM performed a docking maneuver, and at 21:34:00, the two spacecraft docked. Collins opened the CSM hatch and greeted his crewmates before the trio began transferring equipment and sample boxes from the LM to the CSM.

The ascent stage was jettisoned in lunar orbit at 00:01:01 UTC on 22 July. Just before the Apollo 12 flight, it was noted that *Eagle*'s ascent stage was still orbiting the Moon; later NASA reports indicated *Eagle* impacted an "uncertain location" on the lunar surface. The trans-Earth injection burn began at 04:54:42 UTC with a 2.5-minute firing of the SPS engine.

LM-5, Eagle, after undocking from the CSM on the way to the surface. The landing probes may be seen in their extended position on three of the four legs; the plus-Z probe has clearly been removed from under the ladder. (NASA)

CSM-107, Columbia, over Craters Taruntius K, Taruntius P, and Dorsum Cayeux in Mare Fecunditatis. The quad S-Band communications antenna is at the upper left on the SM. The docking port at the apex of the capsule shows up well here. (NASA)

View of the surface from an LM window just after landing, with the LM shadow clearly visible. The descent engine kicked dust out horizontally, essentially parallel to the surface. At engine stop, the flying dust instantaneously disappeared. The crew later mounted a TV camera in this window to photograph their surface activity. (NASA)

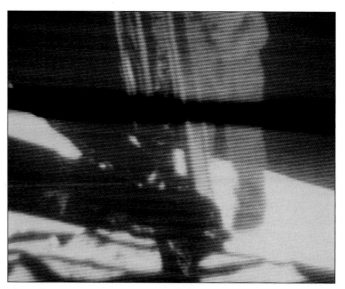

The television camera mounted on the MESA captured this black and white frame of Neil Armstrong descending the ladder of the LM prior to making the first step by man on another celestial body. The black bar running through the center of the picture is an anomaly in the ground data system at the Goldstone Tracking Station. (NASA)

From left to right are spacecraft communicators Charles M. Duke Jr., James A. Lovell, Jr., and Fred W. Haise, Jr., during the lunar landing Lovell had already been to the Moon on Apollo 8, and all three men would get to go to the Moon after Apollo 11. (NASA)

Interior view of the Mission Operations Control Room (MOCR) in the Mission Control Center (MCC) in Building 30 at the MSC during the Apollo 11 lunar extravehicular activity (EVA). The television monitor shows Armstrong and Aldrin on the surface of the Moon. (NASA)

Above, Armstrong collects the contingency lunar surface sample then uses a clothesline device to lift it up to Aldrin in the LM. In the series below, Armstrong turns toward the LM with his face visible through the visor, then lowers his visor, and finally carries the MESA television camera to a different location. All of these frames are from the 16mm Data Acquisition Camera (DAC) mounted in the LM window. (NASA)

HERE MEN FROM THE PLANET EARTH
FIRST SET FOOT UPON THE MOON
JULY 1969, A. D.
WE CAME IN PEACE FOR ALL MANKIND

NEIL A. ARMSTRONG
ASTRONAUT

MICHAEL COLLINS
ASTRONAUT

EDWIN E. ALDRIN, JR.
ASTRONAUT

RICHARD NIXON
PRESIDENT, UNITED STATES OF AMERICA

Close-up of the plaque (left) that was attached to the landing gear strut on the descent stage of the LM. (NASA)

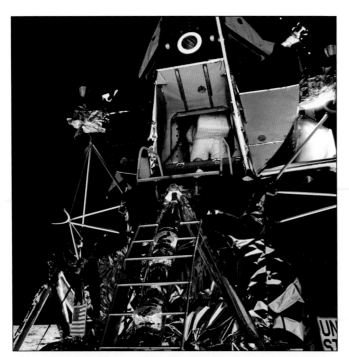

Aldrin descending to the surface. The American flag the astronauts would deploy later is stowed in the long, thin canister attached to the underside of the left-hand rail of the ladder. The black crosses in surface photographs show the level position of the 70mm Hasselblad camera. (NASA)

Aldrin photographing his boot on the lunar surface. Aldrin removed the 70mm Hasselblad camera from the RCU bracket on his chest and took a series of hand-held bootprint pictures. (NASA)

The Apollo 11 Moon Landing

Aldrin made this bootprint on a pristine surface so that he could then photograph it for study by soil mechanics experts. The tread pattern of the lunar EVA boots is clearly visible. (NASA)

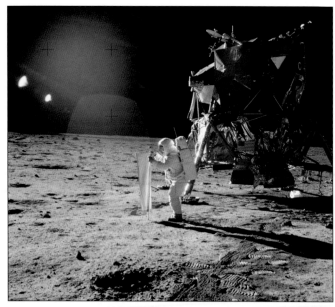

Aldrin erecting the Solar Wind Composition (SWC) experiment, a foil sheet which he pointed toward the Sun. Note the word "shade" printed on the bottom of the collector. (NASA)

Aldrin posing with the SWC experiment. At the end of the EVA, after leaving the SWC exposed for 1 hour and 17 minutes, Aldrin rolled-up the foil and packed it in a bag for analysis on Earth. (NASA)

Aldrin took this photo of the area under the descent stage to document the effects of the engine plume. A radial pattern of scouring is readily visible. Note the gouge made by the probe hanging down from the minus-Y (south) strut at contact. (NASA)

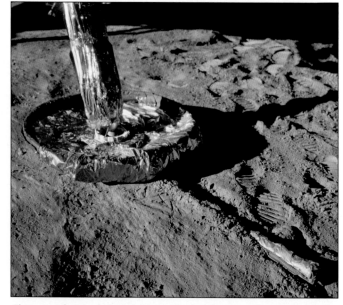

Close-up of one of the LM footpads showing how the contact probe bent outward as the LM landed. Note the different materials used on the engine-facing footpad surfaces as compared with the outer surfaces. (NASA photos, composite by Ron Stephano)

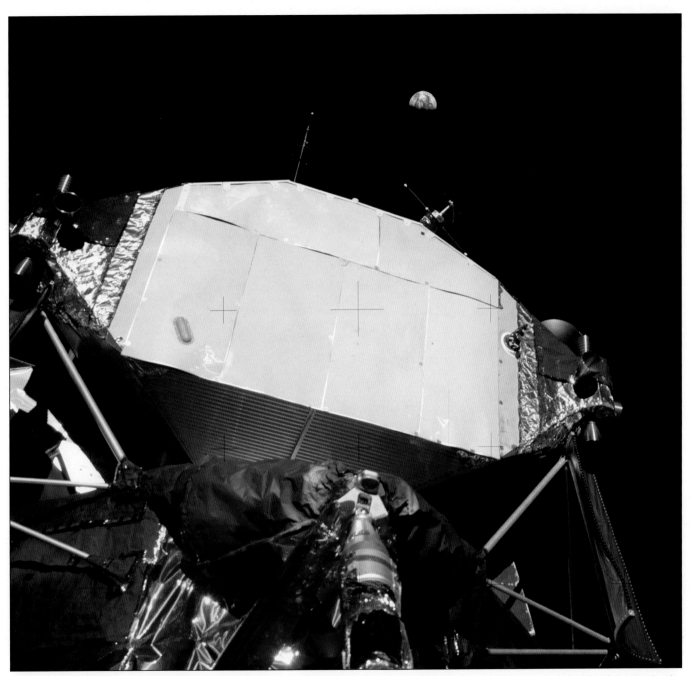

Earth over the LM ascent stage, taken from a spot near the minus-Z (east) footpad. Australia is in sunlight on the left side of the Earth. This positioning explains why the lunar EVA video was received first by the stations at Honeysuckle Creek and Parkes. Oddly, there is some question about who took this photo. With the Earth nearly overhead, Aldrin had to remove the camera from his suit; it is uncertain if he could get the necessary sightline or if he handed the camera to Armstrong to capture this frame and another similar one. (NASA)

The series of small photos were 16mm frames taken from the LM window while Armstrong and Aldrin set up the American flag on the surface. The larger photo is a post-EVA photo taken from Aldrin's LM window that shows the flag, television camera, and the cluster of boulders beyond which were probably ejected from West Crater. Note the raised rim on the fresh, young crater in the middle distance beyond and to the right of the television camera. Note, also, the cable running from the MESA to the television camera. The difference between the darker, heavily disturbed soil around the camera, and the undisturbed light soil where the astronauts did not set foot is clearly visible. Jack Schmitt speculates that the descent engine plume swept away the smallest particles, leaving a higher than normal percentage of larger particles. The more jagged surface is a better reflector of sunlight than the normal surface and, therefore, appears brighter. (NASA)

Aldrin salutes the flag; his fingertips are visible on the far side of his faceplate. Note the well-defined footprints in the foreground. (NASA)

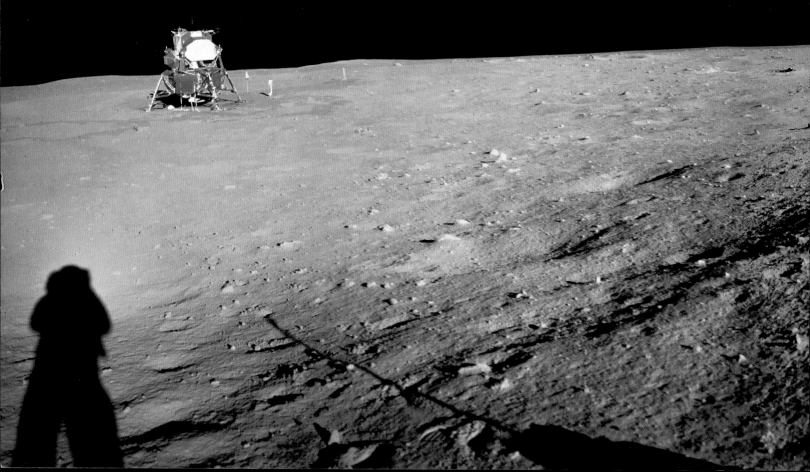

Composite of several 70mm Hasselblad images showing the LM and East Crater. Armstrong walked about 200 feet east of the LM to look at the 100-foot diameter crater. The handle of the Gold camera is visible in the center of the photo. (NASA photos, composite by Ron Stephano)

The beautiful 70mm Hasselblad images were taken on color film, but in 1969 there was no available technology to downlink similar images. Instead, the world watched the lunar EVA on grainy, black and white television. These photographs were made from a televised image received at the Deep Space Network tracking station at Goldstone, California. In the upper left image, President Richard M. Nixon had just spoken to the astronauts by radio and Aldrin, a colonel in the U.S. Air Force, is saluting the president. (NASA)

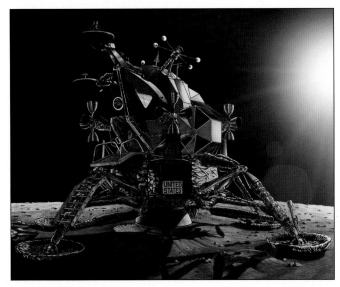

This caricature of the Apollo 11 Lunar Module by our friend, Ron Stephano, shows amazing, if distorted, detail. (©2009 Ron Stephano)

On facing page: *No other Apollo photograph has been reproduced as often as this portrait of Aldrin. Armstrong's reflection is visible in the visor. Aldrin has his left arm raised and is probably reading the checklist sewn on the cuff of his glove. He is moving his right foot forward, as can be seen by the mound of dirt building up in front of the toe of that boot. Note the dirt adhering to Buzz's boots and knees. Otherwise, he is remarkably clean. Oddly, however, this frame (AS11-40-5903) was manipulated before it was released immediately after the mission (since this was a 70mm Hasselblad image on color film). A close examination shows that the OPS antenna on Aldrin's helmet is missing. The explanation is simple – Armstrong did not get the top of the helmet in the original frame. NASA public affairs retouched the image, adding the last couple of inches of the helmet, PLSS, and black sky. The retouched image was provided to Life Magazine, and was also used in several NASA publications (EP-72, Log of Apollo 11; and EP-73, The First Lunar Landing as Told By the Astronauts), published immediately after the mission.* (NASA)

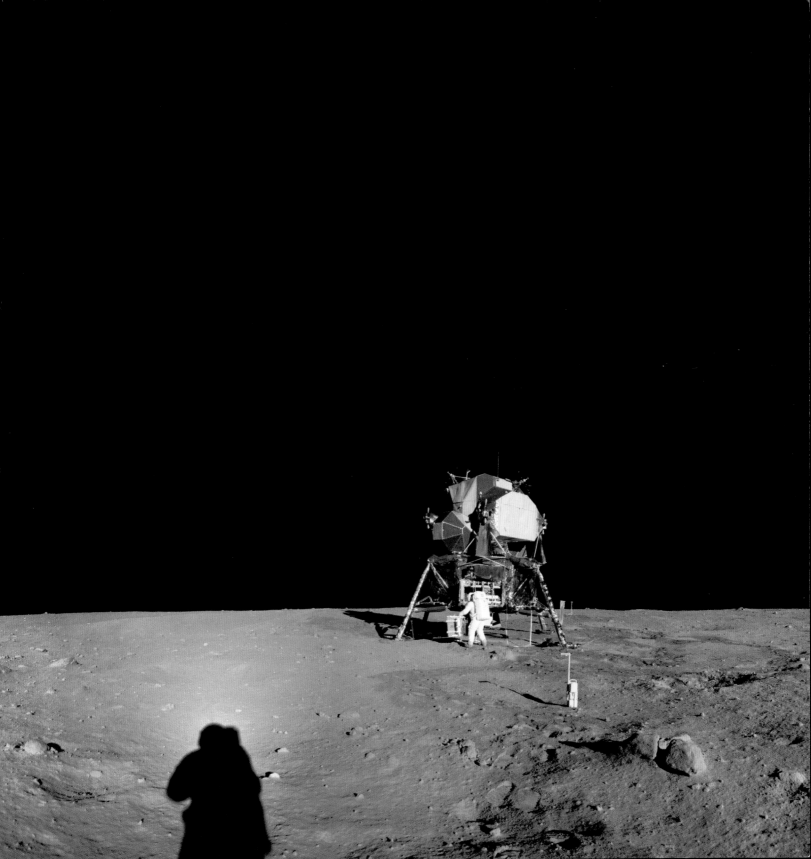

This is the first in a sequence of photographs that Armstrong took of Aldrin carrying the EASEP to the deployment site south of the spacecraft. Aldrin is walking across the raised rim of a small crater. The Lunar Ranging Retro Reflectometer (LRRR) is in his right hand while the seismometer package is in his left hand. (NASA)

Aldrin carried the equipment to an intermediate point away from the LM. At this point, Aldrin was responsible for finding a good location for the seismometer and setting it up; Armstrong was responsible for finding a location for the LRRR and setting it up. Note the depth of the footprints Aldrin has left in the soft rim of the crater. (NASA)

Aldrin leveling the seismometer after the west-side solar array "deployed automatically"; the east-side array has not yet deployed. Note the baton-like transmitting antenna and the three-pronged ("cricket wicket") gnomon on the top of the main package. (NASA)

Both seismometer solar arrays are now deployed. Aldrin is looking toward the LM, perhaps to get a reference for his alignment. The Gold camera is at the extreme right edge of the photo between Aldrin and the spacecraft. (NASA)

Aldrin deploying the Passive Seismic Experiments Package (PSEP). The Laser Ranging Retro-Reflector (LRRR) can be seen to the left and further in the background. In the far left background is the black and white television camera and American flag. (NASA)

Post-deployment photograph of the LRRR with the Gold camera perched on a rock in the background. The retroreflectors are still being used by the University of Texas McDonald Observatory, 40 miles west of Austin in west Texas. (NASA)

Armstrong took this image while he advanced the film prior to removing the magazine. The view to the northwest shows the television camera and the Solar Wind Composition (SWC) experiment. The dark area at lower left is probably part of Armstrong's suit. (NASA)

Aldin driving the first of two core tubes into the surface, although he only managed to get it in about 8 inches. The Solar Wind Composition experiment is just beyond the core tube and the television camera is at the extreme left. (NASA)

Excellent composite of several 70mm Hasselblad images showing the view from a LM window. (NASA photos, composite by Ron Stephano)

Earthrise as seen by Mike Collins in the Command Module on 20 July 1969. When these photos were taken, Columbia was passing over Mare Smythiion on revolution (orbit) 12 or 13. (NASA)

The Apollo 11 Moon Landing

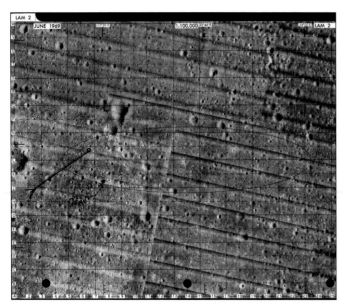

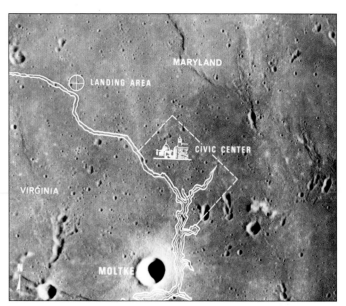

Mike Collins used this map to mark the estimated LM locations provided by Houston. The actual landing site was about 750 feet from the arrow labeled "Last Bst Pos Prior L/O." (Scan courtesy Bob Craddock and Alan Needell, National Air and Space Museum)

A photographic illustration comparing the size of Apollo Landing Site 2 with metropolitan Washington, D.C. The white overlay is printed over a lunar surface photograph taken from Apollo 10 during its lunar orbit mission. (NASA)

Above and facing page: *The LM ascent stage seen by Collins in the CM as it approached from below. At right, the maze area in the background is Mare Smythii. According to his autobiography, Collins never felt lonely during his day of solo flying around the Moon. Although it had been said that "not since Adam has any human known such solitude," Collins felt very much a part of the mission and wrote, "this venture has been structured for three men, and I consider my third to be as necessary as either of the other two." During the 48 minutes of each orbit that he was out of radio contact with Earth, Collins felt "awareness, anticipation, satisfaction, confidence, almost exultation." (NASA)*

RETURNING HEROES

Once the SPS engine had fired and established them on a trajectory toward Earth, the astronauts mostly just waited. Mission Control mentioned that all but four nations – Albania, China, North Korea, and North Vietnam – had told their citizens about the Moon landing, and most had carried live television coverage of the event.

On 23 July, the astronauts made a television broadcast on the last night before splashdown. Mike Collins commented, "The Saturn V rocket which put us in orbit is an incredibly complicated piece of machinery, every piece of which worked flawlessly … We have always had confidence that this equipment will work properly. All this is possible only through the blood, sweat, and tears of a number of a people … All you see is the three of us, but beneath the surface are thousands and thousands of others, and to all of those, I would like to say, 'Thank you very much.'"

Buzz Aldrin said, "This has been far more than three men on a mission to the Moon; more, still, than the efforts of a government and industry team; more, even, than the efforts of one nation. We feel that this stands as a symbol of the insatiable curiosity of all mankind to explore the unknown … Personally, in reflecting on the events of the past several days, a verse from Psalms comes to mind. 'When I consider the heavens, the work of Thy fingers, the Moon and the stars, which Thou hast ordained; What is man that Thou art mindful of him?'"

Armstrong concluded, "The responsibility for this flight lies first with history and with the giants of science who have preceded this effort … We would like to give special thanks to all those Americans who built the spacecraft; who did the construction, design, the tests, and put their hearts and all their abilities into those craft. To those people tonight, we give a special thank you, and to all the other people that are listening and watching tonight, God bless you. Good night from Apollo 11."

As the CM approached Earth, the weather in the primary recovery area (1,475 miles downrange from the entry interface of 400,000 feet) in the Pacific Ocean began to worsen. Mostly because of building thunderstorms, the targeted landing point was moved to 1,725 miles from entry interface, meaning Collins would need to fly 250 miles further downrange. As they approached Earth, the crew jettisoned the SM at 16:21:13 UTC on 24 July and turned the CM so the heat shield pointed in the direction of travel. Returning from the Moon meant the capsule would enter the Earth's atmosphere at 36,194 feet per second (24,677 mph), and the crew would be subjected to a deceleration of 6.73-g.

When the CM hit the entry interface, Aldrin triggered a camera to capture the colors around the plasma sheath – lavenders, little touches of violet, and great variations of blues and greens wrapped around an orange-yellow core. A surprisingly small amount of material seemed to be flaking off the heat shield; Collins did not see the chunks he had seen in Gemini.

The forward heat shield over the CM was jettisoned at about 24,000 feet, followed 1.5 seconds later by the two drogue parachutes that oriented the capsule properly and provided initial deceleration. At about 10,000 feet, the drogue chutes were released and the three pilot chutes deployed; these pulled the main parachutes from the forward section of the CM. The main parachutes held the CM at an angle of 27.5 degrees so the capsule hit the water on its "toe," providing water penetration with the least impact load.

The recovery area had a visibility of 14 miles and winds of 18 mph with 3-foot waves. Search aircraft spotted the capsule almost 11 minutes before splashdown, and radar aboard the aircraft carrier USS *Hornet* (CVS-12) established contact a minute later. *Columbia* splashed down at 16:50:35 UTC (12:50:35 p.m. EDT) on 24 July 1969, after a mission elapsed time of 195 hours, 18 minutes, 35 seconds. The splashdown point was 13 degrees, 19 minutes north, 169 degrees, 9 minutes west, about 210 miles south of Johnston Atoll and 15 miles from the *Hornet*.

Upon landing, surface winds filled the parachutes and flipped *Columbia* on its nose in the water, but floatation bags righted the capsule in less than 8 minutes. After the CM was righted, four biological isolation garments and decontamination gear were lowered from a helicopter to one of the nearby rafts where one of four rescue swimmers donned a biological isolation garment. The swimmer then moved the raft near the CM, the crew opened the hatch, the swimmer tossed in three biological isolation garments, and the crew closed the hatch.

After donning the biological isolation garments, the crew egressed at 17:21 UTC and the protected swimmer sprayed the upper deck and hatch areas with Betadine, a water-soluble iodine solution. After the four men and the raft were wiped with a solution of sodium hypochlorite, they were retrieved by helicopter and taken to the *Hornet*, arriving 63 minutes after splashdown. Apollo 11 had traveled 828,743 miles, including an excursion 210,391 miles from Earth. It had been a long trip.

View of Earth from Apollo 11 prior to entry. Note the dramatic red tint of the area near the terminator (the dividing line between the illuminated and the unilluminated areas). (NASA)

Quarantine

Strict quarantine procedures had been developed due to fears that the Moon might contain undiscovered pathogens. After the helicopter landed on the *Hornet*, it was moved the hangar deck next to the Mobile Quarantine Facility. The astronauts walked across the deck and entered the trailer along with a physician and medical technician. The CM was moved onto the *Hornet* approximately 3 hours after landing and was attached to the Mobile Quarantine Facility through a flexible tunnel, and the lunar surface samples, film, data tape, and medical samples were removed from the CM. Once the quarantine facility was secured, the rescue swimmers, helicopter, rafts, and the deck of the *Hornet* were disinfected with either Betadine or a glutaraldehyde solution.

President Richard M. Nixon was aboard the *Hornet* to welcome the astronauts back to Earth.

The CM and Mobile Quarantine Facility were off-loaded from the *Hornet* in Hawaii at 00:15 UTC on 27 July. The Mobile Quarantine Facility was loaded aboard an Air Force Lockheed C-141A Starlifter and flown to the Lunar Receiving Laboratory in Houston, where it arrived at 06:00 on 28 July. The safing of the CM pyrotechnics was completed at 02:05 on

The lunar missions provided dramatic views of Earth from a distance, such as this photo from Apollo 11 showing one-third of the Earth's sphere illuminated. (NASA)

27 July, after which the CM was flown on a Douglas C-133 Cargomaster to Houston, where it arrived at 23:17 on 30 July.

Twenty persons on the medical support teams were exposed, directly or indirectly, to lunar material for periods ranging from 5 to 18 days. Daily medical observations and periodic laboratory examinations showed no signs or symptoms of infectious disease related to lunar exposure. No microbial growth was observed from the prime lunar samples after 156 hours of incubation on all types of differential media. No microorganisms that could be attributed to an extraterrestrial source were recovered from the crewmen or the spacecraft. None of the 24 mice injected with lunar material showed visible shock reaction following injection, and all remained alive and healthy during the first 10 days of a 50-day toxicity test. During the first 7 days of testing of the prime lunar samples in germ-free mice, all findings were consistent with the decision to release the crew from quarantine.

Apollo 11 Epilogue

On 10 August, the astronauts exited quarantine to the cheers of the American public. Over the next couple of weeks, parades were held in their honor in New York, Chicago, Los Angeles, and Mexico City.

In Los Angeles, there was an official State Dinner on 13 August to celebrate Apollo 11, attended by Members of Congress, 44 governors, the Chief Justice, and ambassadors from 83 nations. President Richard Nixon and Vice President Spiro T. Agnew honored each astronaut with a presentation of the Presidential Medal of Freedom. This celebration was the beginning of a 45-day "Giant Leap" tour that brought the astronauts to 25 foreign countries and included visits with prominent leaders such as Queen Elizabeth II of the United Kingdom. Many nations would honor the first manned Moon landing by issuing Apollo 11 commemorative postage stamps or coins.

On 16 September 1969, the three astronauts spoke before a Joint Session of Congress, and presented two U.S. flags, one to the House of Representatives and the other to the Senate, that had been carried to the surface of the Moon with them.

After it was decontaminated with formaldehyde, the CM was shipped inside an Aerospacelines B377SG Super Guppy aircraft to the North American Aviation facility in Downey, California, on 14 August for post-flight analysis. At the conclusion of the engineering work, the capsule was transferred to the Smithsonian. The CM, the only part of the spacecraft that returned to Earth, is displayed in the Milestones of Flight Gallery at the National Air and Space Museum in Washington, D.C. The quarantine trailer is displayed at the Udvar-Hazy Center near Washington Dulles International Airport in Virginia.

Ultimately, the Apollo Program returned 841.5 pounds of material from the Moon, much of which is stored at the Lunar Receiving Laboratory in Houston. In general, the rocks are extremely old compared to rocks found on Earth. As measured by radiometric dating techniques, they range from about 3.2 billion years old for the basaltic samples from the lunar mare, to about 4.6 billion years for samples from the highlands crust. As such, they represent samples from a very early period in the development of the Solar System that is largely missing from Earth. A particularly important sample, called the Genesis Rock, was retrieved by James Irwin and David Scott during the Apollo 15 mission. This rock, called anorthosite, is composed almost exclusively of the calcium-rich feldspar mineral anorthite, and is believed to be representative of the highland crust. A geochemical component called KREEP was discovered that has no known terrestrial counterpart. Together, KREEP and the anorthositic samples have been used to infer that the outer portion of the Moon was once molten.

Almost all of the rocks show evidence for having been affected by impact processes. For instance, many samples appear to be pitted with micrometeoroid impact craters, something that is never seen on Earth due to its thick atmosphere. Additionally, many show signs of being subjected to high-pressure shock waves that are generated during impact events. Some of the returned samples are of impact melt, referring to materials that are melted near an impact crater. Finally, all samples returned from the Moon are highly brecciated as a result of being subjected to multiple impact events.

Analysis of composition of the lunar samples led to the conclusion, reached in 1984, that the Moon was created through a "giant impact" of a large astronomical body with the Earth.

Program Costs

Ultimately, the Apollo program included 15 manned missions: six Earth orbiting missions (Apollo 7 and 9, three manned Skylab missions, and Apollo-Soyuz), two lunar orbiting missions (Apollo 8 and 10), an unplanned lunar swingby (Apollo 13), and six lunar landing missions (Apollo 11, 12, 14, 15, 16, and 17). Two astronauts from each of these six missions walked on the Moon – Neil A. Armstrong, Edwin E. Aldrin, Jr., Charles Conrad, Jr., Alan L. Bean, Alan B. Shepard, Jr., Edgar D. Mitchell, David R. Scott, James B. Irwin, John W. Young, Charles M. Duke, Jr., Eugene A. Cernan, and Harrison H. Schmitt; they are the only humans to have set foot on another celestial body.

Twelve other astronauts flew to the vicinity of the Moon – Frank F. Borman II, James A. Lovell, William A. Anders, Thomas P. Stafford, Michael Collins, Richard F. Gordon, Jr., John L. "Jack" Swigert, Fred W. Haise, Jr., Stuart A. Roosa, Alfred M. Worden, Thomas K. "Ken" Mattingly II, and Ronald E. Evans, Jr.

Cernan, Lovell, and Young are the only men to fly more than one lunar mission, logging two each.

The cost of the Apollo program is estimated at $125 billion (2009 dollars, or $25.4 billion in 1969 dollars). The CSM cost $17 billion to develop, the LM $11 billion, and the Saturn boosters another $46 billion. The remaining costs were associated with building the launch site, mission control, simulation facilities, test facilities, and other infrastructure needed for the program.

Originally three additional lunar landing missions had been planned, Apollo 18 through Apollo 20. These missions were cancelled to make funds available for the development of the Space Shuttle, and to make their vehicles available to the Skylab program. Only one of the three remaining Saturn Vs was actually used; the others became museum exhibits.

This six-inch-long gold olive branch, the traditional symbol of peace, was left on the Moon's surface by the Apollo 11 crew. They also left a silicon disk that had messages from 77 world leaders. (NASA)

A Sikorsky SH-3D Sea King from U.S. Navy squadron HS-4 hovers above Apollo 11 seconds after it splashed down in the Pacific Ocean. The spacecraft settled into the Stable II Position (apex down), as shown here, but inflatable floatation bags quickly righted it. (NASA)

U.S. Navy underwater demolition team (UDT) swimmer Lt. Clancy Hatleberg (without life vest) in the decontamination raft with the Apollo 11 crew during recovery operations. Earlier, Hatleberg sprayed decontamination fluid on the crew and capsule. (NASA)

The once pristine capsule was severely charred during its 24,677-mph entry into Earth's atmosphere, but the heat shield performed as-expected. The orange floatation collar was attached by Navy divers to stabilize the capsule while the crew egressed. (NASA)

The Command Module being lowered to the deck of the U.S.S. Hornet (CVS-12). The flotation collar attached by Navy divers is being removed from the capsule, and the floatation bags can the capsule are beginning to deflate through normal leakage. (NASA)

After the CM was righted into the Stable I Position (apex up), four biological isolation garments and decontamination gear were lowered from a helicopter to one of the nearby rafts where one of four rescue swimmers donned a biological isolation garment. The swimmer, Lt. Clancy Hatleberg, then moved the raft near the CM, the hatch was opened, three biological isolation garments were tossed into the CM, and the hatch was closed. After donning the biological isolation garments, the crew egressed at 17:21 and the protected swimmer sprayed the upper deck and hatch areas with Betadine, a water-soluble iodine solution. After the four men and the raft were wiped with a solution of sodium hypochlorite, they were retrieved by helicopter and taken to the Hornet, arriving 63 minutes after splashdown. (NASA)

Above and two below: *Various views of the Mission Operations Control Room (MOCR) in the Mission Control Center (MCC) after the Apollo 11 crew had been recovered. The television monitor in the top photo shows President Richard M. Nixon greeting the astronauts aboard the Hornet.* (NASA)

Even NASA watched Walter Cronkite, shown on the monitors in the MOCR. Three Apollo principals, (from left), Dr. George M. Low, Manager of the Apollo Spacecraft Program Office, Dr. George E. Mueller, Associate Administrator for Manned Space Flight, and Dr. Charles A. Berry, MSC flight surgeon, celebrate. (NASA)

The Apollo 11 crew, wearing biological isolation garments, arrive aboard the Hornet and walk toward the Mobile Quarantine Facility (MQF) where they will be confined until they arrive at the Lunar Receiving Laboratory (LRL) in Houston. (NASA)

Two hours after the crew, the capsule was hoisted aboard the Hornet and positioned near the MQF. A plastic tunnel was connected between the hatch and the MQF door, and the moon rocks, space suits, and cameras were moved into the MQF. (NASA)

President Richard M. Nixon was aboard the Hornet to greet the Apollo 11 crew after they were confined in the Mobile Quarantine Facility. From left are Armstrong, Collins, and Aldrin. (NASA)

The traditional post-flight cake cutting ceremony was altered somewhat since the Apollo 11 crew were in isolation inside the Mobile Quarantine Facility. Nevertheless, the ceremony contained the pomp and circumstance deserving of the occasion. (NASA)

From left, Collins, Aldrin, and Armstrong relax in the Mobile Quarantine Facility. The crew spent two-and-one-half days in the Mobile Quarantine Facility enroute from the Hornet to the Lunar Receiving Laboratory in Houston. (NASA)

The Apollo 11 Mobile Quarantine Facility was off-loaded from the Hornet at Pearl Harbor and sent to nearby Hickam AFB to be airlifted to Houston aboard an Air Force Lockheed C-141 Starlifter transport. This was actually the third Mobile Quarantine Facility (MQF003) manufactured; MQF001 was intended, but never used, for Apollo 13, MQF002 for Apollo 12, and MQF004 for Apollo 14. Based on the results of these flights, subsequent Apollo missions did not subject the crew to quarantine. The Mobile Quarantine Facility was a specially built Airstream trailer that contained living and sleeping quarters, a kitchen, a latrine, and own internal backup power, air conditioning, and communications systems. Quarantine was assured by maintaining negative internal pressure. (NASA)

The Apollo 11 Moon Landing

The Mobile Quarantine Facility with the Apollo 11 crew inside, is unloaded from the C-141 at Ellington Field after a flight from Hawaii. The crew remained in the MQF until they arrived at the Crew Reception Area of the Lunar Receiving Laboratory. (NASA)

The Apollo 11 wives, (from left) Pat Collins, Jan Armstrong, and Joan Aldrin, greet the crew after they arrive at Ellington Field. Looking through the Mobile Quarantine Facility window are (from left) Armstrong, Aldrin, and Collins. (NASA)

The Apollo 11 crew after arriving at Ellington Field. Armstrong is strumming on a ukulele while Collins (right foreground) and Aldrin are looking out the window. In the picture (at left) are MQF support personnel. A large crowd was present to welcome the astronauts home to Houston. Once inside the MQF, the cumbersome biological isolation garments used during recovery were replaced with much more comfortable NASA flight suits for the rest of the quarantine period. (NASA)

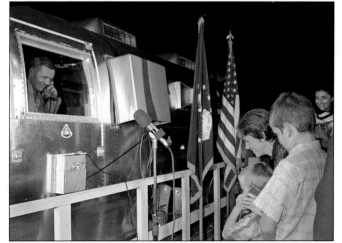

Neil Armstrong greets his son Mark on the telephone intercom system, while his wife Jan and other son Eric look on. For the Apollo 11 crew , the MQF was home for 65 hours after splashdown, including the sea voyage to Pearl Harbor and the nine-hour flight to Houston. Quarantined in the MQF with the astronauts were Apollo 11 flight surgeon Dr. William Carpentier and technician John Hirasaki. The returning astronauts were scrutinized during their entire stay, undergoing various medical tests and examinations. (NASA)

LUNAR RECEIVING LAB

The Lunar Receiving Laboratory (LRL) was constructed to quarantine astronauts and lunar material to mitigate the risk of contaminating the Earth by unknown organisms, and also to eliminate the possibility of back-contamination of the lunar samples. Construction of the laboratory was complete by the middle of 1967, although equipment installation continued for another year. The quarantine of the Apollo 11 crew was uneventful, and no signs or symptoms of infectious disease related to lunar exposure became apparent in any of the crewmen or support staff. The lunar samples were analyzed using glove-box vacuum chambers, and no microorganisms attributable to an extraterrestrial source were recovered from the crewmen or the spacecraft. Release of crew, equipment, and lunar samples took place on schedule, and the quarantine requirement was dropped beginning with Apollo 15. The LRL is now used for study, distribution, and storage of the lunar samples. In 1976 some samples were moved to Brooks AFB in San Antonio for second-site storage.

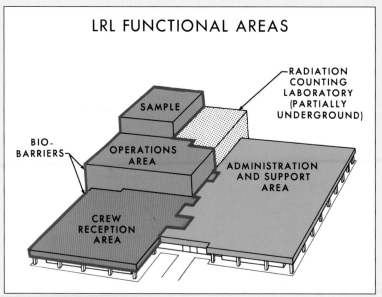

Artist's concept of the Lunar Receiving Laboratory (LRL) in Building 37 at the Manned Spacecraft Center showing the three major functional areas. The Crew Reception Area contained the astronaut living quarters during the quarantine period. (NASA)

Typical astronaut quarters in the Crew Reception Area of Lunar receiving Laboratory. The rooms, like the rest of the facility, were simple but comfortable. The Apollo 11 crew arrived in the LRL on 28 July and were released from quarantine on 10 August. (NASA)

Astronaut debriefing and press conferences took place in Room 190 of the Administration and Support Area of the Lunar Receiving Laboratory. The astronauts were isolated on the far side of the glass wall. The chairs do not appear to be particularly comfortable. (NASA)

The Apollo 11 astronauts eating dinner in the Crew Reception Area of the Lunar Receiving Laboratory. From left are Buzz Aldrin, Mike Collins, and Neil Armstrong. (NASA)

The crew of Apollo 11 at a postflight debriefing session. In the foreground are Donald K. Slayton (right), MSC Director of Flight Crew Operations, and Lloyd Reeder, training coordinator. (NASA)

Oddly, given the small number of guests, meals were served cafeteria style in the Crew Reception Area of the Lunar Receiving Laboratory. From left are Buzz Aldrin, Mike Collins, and Neil Armstrong. (NASA)

On 5 August 1969, Neil Armstrong celebrated his 39th birthday in the Lunar Receiving Laboratory. The staff prepared a birthday cake, complete with a Lunar Module decoration. At this point, Armstrong was arguably the most famous person in the world. (NASA)

At left is the first Apollo 11 sample return container arriving at Ellington Field. Posing with the container are, from left, Dr. George M. Low, Manager of the Apollo Spacecraft Program Office; Lt. Gen. Samuel C. Phillips, Apollo Program Director, George S. Trimble, MSC Deputy Director (mostly obscured); Eugene G. Edmonds, MSC Photographic Technology Laboratory; Dr. Richard S. Johnston (in back), Special Assistant to the MSC Director; Dr. Thomas O. Paine, NASA Administrator; and Dr. Robert R. Gilruth, MSC Director. At right, the same sample return container arrives at the Lunar Receiving Laboratory a little while later. (NASA)

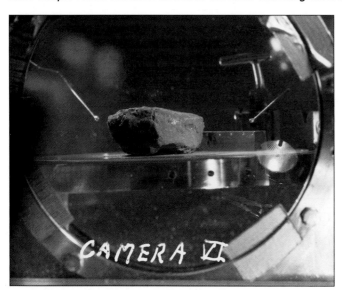

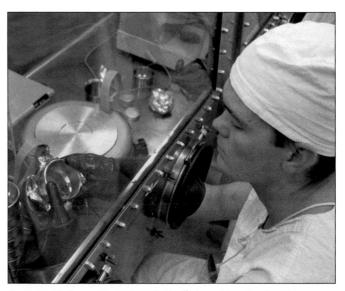

This is the first lunar sample (10003) examined in detail in the Lunar Receiving Laboratory, a granular, fine-grained, mafic (iron magnesium rich) rock. The scale was printed backwards due to the photographic configuration in the vacuum chamber. (NASA)

Dr. Grant Heikan, a Lunar Sample Preliminary Examination Team member, examines lunar material in a sieve from the sample return container which was opened in the biopreparation laboratory of the Lunar Receiving Laboratory. (NASA)

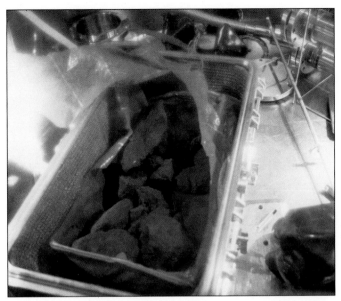

This is the Documented Sample Apollo Lunar Sample Return Container and contains approximately 20 grab samples weighing a total of 12 pounds. Armstrong also packed the two core tubes in this box and these can be seen at the upper left. The gloved right hand of the vacuum-chamber operator is at the lower right. (NASA)

The second Apollo 11 sample return container in the vacuum laboratory of the Lunar Receiving Laboratory. This container was opened for the first time at 1 pm (CDT) on 4 August 1969. Some of the material has already been removed from the container and placed in the stainless steel can in the photo at left. (NASA)

Dr. Ross Taylor (seated), Australian National University, and John Allen, Brown and Root-Northrop technician, review preliminary data from the optical emission spectrograph in the spectrographic laboratory of the Lunar Receiving Laboratory. (NASA)

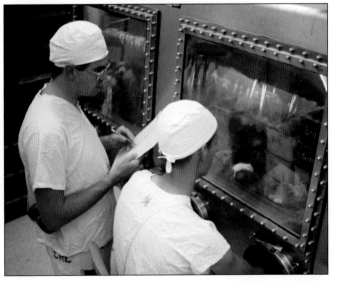

Landrum Young (seated) of Brown and Root-Northrop, and Russell Stullken of NASA, examine mice which have been inoculated with lunar sample material in the animal laboratory. The mice exhibited no unusual behavior or symptoms during the experiments. (NASA)

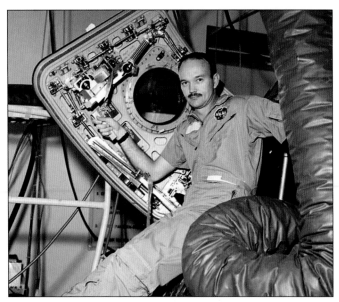

Mike Collins sits in the Command Module hatch after its return to the Lunar Receiving Laboratory for detailed examination. The large brown tube is blowing cooling air into the capsule. (NASA photo scanned by J.L. Pickering)

The Apollo 11 Command Module is loaded aboard a Boeing-Aerospacelines B377SG Super Guppy at Ellington Field for shipment to North American Aviation in Downey, California. The CM was just released from its post-flight quarantine at the MSC. (NASA)

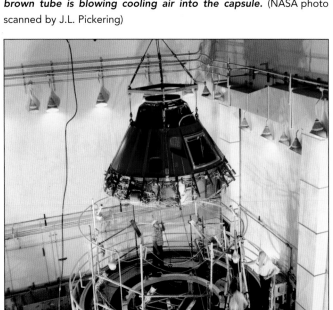

The Command Module is prepared for a "vacuum bake" in the thermal vacuum chamber at Downey on 5 September 1969. This was intended to bake any residual moisture from the splashdown before being encased in plastic for display at the Smithsonian. (Rockwell)

The Apollo 11 Command Module is on display in the Milestones of Flight gallery at the Smithsonian National Air and Space Museum in Washington, D.C. The forward portion of Columbia's heatshield was discarded during reentry. (Dennis R. Jenkins)

New York City welcomed the Apollo 11 crew with a showering of ticker tape down Broadway and Park Avenue in a parade described as the largest in the city's history. Pictured in the lead car, from the right, are Neil Armstrong, Mike Collins, and Buzz Aldrin. (NASA)

Chicago Also threw a ticker-tape parade for the crew before they left on a 45-day "Giant Leap" tour that brought the astronauts to 25 foreign countries and included visits with prominent leaders such as Queen Elizabeth II of the United Kingdom. (NARA)

GOING BACK

On 14 January 2004, President George W. Bush announced the Vision for Space Exploration that would return men to the moon for the first time since Apollo 17. Later renamed the Constellation Program (abbreviated CxP), the Vision attempts to recreate and extend what was done in Apollo.

Constellation consists of the Orion spacecraft (originally called the Crew Exploration Vehicle, CEV), and its Ares I (Crew Launch Vehicle, CLV) booster, and the Altair lunar lander (the Lunar Surface Access Module, LSAM) and its Ares V (Cargo Launch Vehicle, CaLV) booster. The Ares I uses a five-segment variant of the Space Shuttle Solid Rocket Booster along with a new J-2X-powered Upper Stage. The Ares V uses six RS-68B engines underneath a large propellant tank, flanked by a pair of 5.5-segment solid rocket boosters and topped by a J-2X-powered Earth Departure Stage (EDS).

The four-seat Orion is split, much like Apollo, into a Crew Module (CM, instead of Command Module) and a Service Module (SM). Plans call for the phased introduction of variants tailored for specific missions. The Block I Orion will be used for International Space Station (ISS) crew rotation, while the Block II variant will be used to return to the Moon.

Like its Apollo predecessor, Altair consists of two parts: an ascent stage that houses the four-person crew, and a descent stage that has the landing gear, the majority of the consumables (oxygen and water), and scientific equipment. Unlike Apollo, Altair will land in the lunar polar regions favored by NASA for future lunar base construction.

Originally, the Ares launch vehicles were supposed to be based largely on existing Space Shuttle components, but have evolved considerably since the beginning of the program and share little with their legacy systems. The J-2X engine for the Ares I Upper Stage is based, loosely, on the Rocketdyne J-2 used in the S-II and S-IVB stages of the Saturn V, while the Rocketdyne RS-68A engine for Ares V is based on the engine used by the Boeing Delta IV Evolved Expendable Launch Vehicle.

Currently the first test flight of the Ares I-X (a four-segment booster with a dummy Upper Stage) is scheduled for the second quarter of 2009, followed by a test flight (Ares I-Y) of a five-segment booster in mid-2013. The first manned flight of Orion is scheduled for 2016. Unlike President Kennedy's "end of the decade" schedule, no concise timeline has been established for returning men to the Moon.

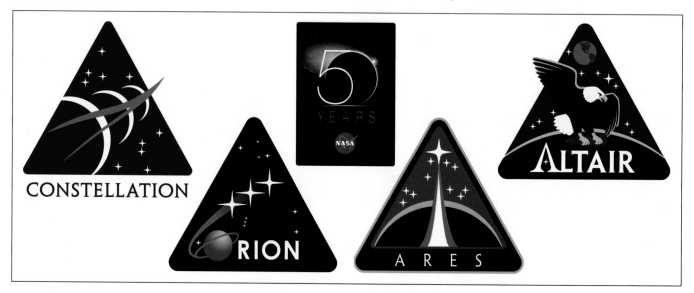

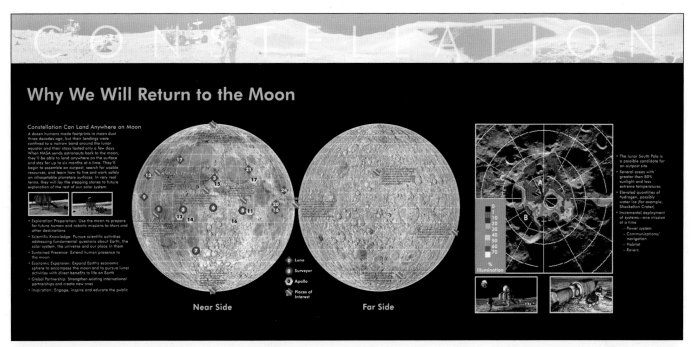

This promotional poster from NASA attempts to explain why we are going back to the Moon, highlighting that Constellation is being designed to land at locations other than the equatorial sites chosen by Apollo. (NASA)

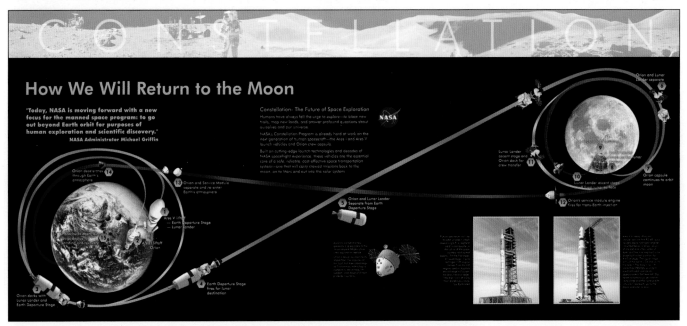

Unlike Apollo, each Constellation mission to the Moon will require two launches. An Ares I Crew Launch Vehicle will loft an Orion spacecraft with a four-person crew into orbit where it will rendezvous with an Altair lunar lander launched by an Ares V Cargo Launch Vehicle. (NASA)

The Orion spacecraft (upper center of the illustration) is being developed as a multi-purpose vehicle. Here, it is shown docked to the International Space Station, which will be its first destination while the development of the Altair and Ares V (below) is completed. (NASA)

The Apollo 11 Moon Landing

This is the stack Constellation plans to send to the Moon. The Earth Departure Stage (EDS) is functionally similar to the S-IVB stage used by the Saturn V. On top of it is an Altair lunar lander docked to an Orion spacecraft with its unique round solar arrays. (NASA)

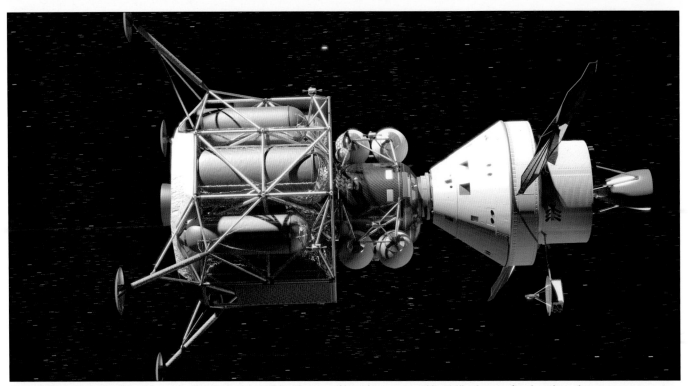

One of the major departures from the Apollo legacy will be the use of liquid oxygen and liquid hydrogen for the Altair descent stage engines instead of the hypergolic propellants used by the LM. This forces the Altair descent stage to be much larger than the LM. (NASA)

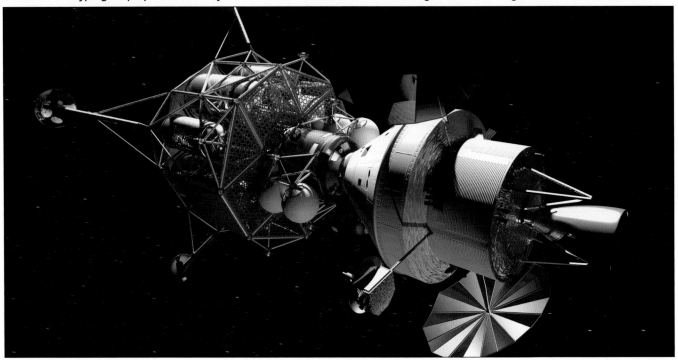

The Apollo 11 Moon Landing

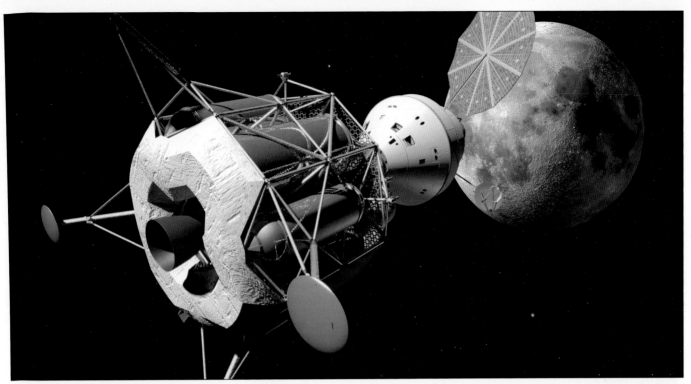

In another major departure from Apollo, all four crewmembers will descend to the surface in the Altair, leaving the Orion unattended in orbit. The height of the Altair means these astronauts will have a much longer climb to the surface than Armstrong and Aldrin. (NASA)

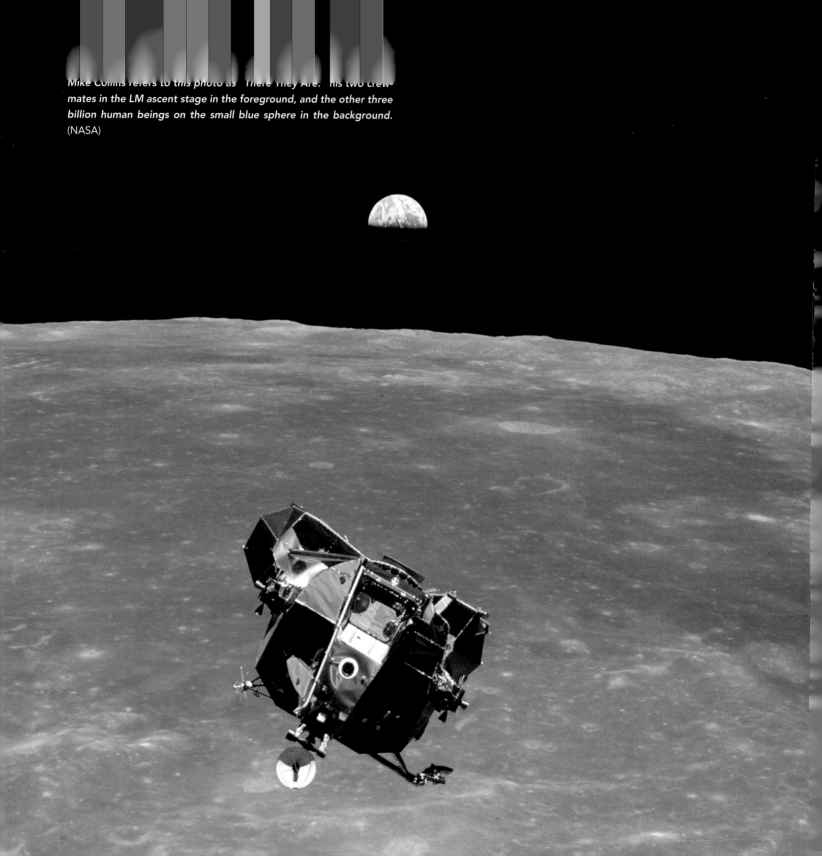

Mike Collins refers to this photo as "There They Are" his two crewmates in the LM ascent stage in the foreground, and the other three
billion human beings on the small blue sphere in the background.
(NASA)